HIDDEN LANDSCAPES

PHOTOMICROGRAPHS OF AN UNSEEN UNIVERSE

STEVEN BROOKE

For Miles and Donna
and for
Dennis Jenkins and Jean Ann Welch

First published in the United States of America in 2016 by
STEVEN BROOKE STUDIOS, INC.
www.stevenbrooke.com

ISBN-13: 978-0-9801212-2-3
Library of Congress Control Number: 2016900339
Printed and bound in the United States of America

To purchase original prints please contact
steven@stevenbrooke.com

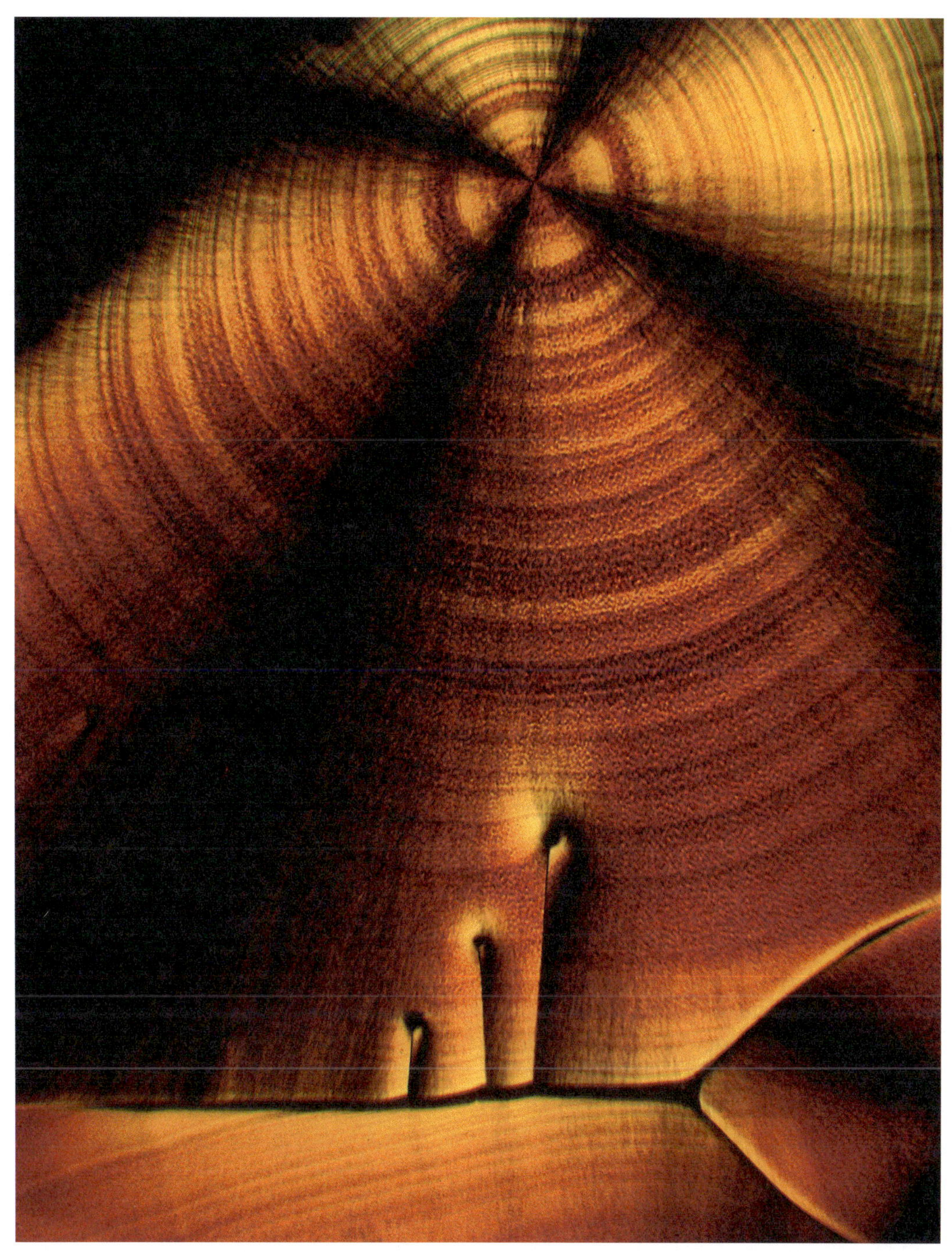

HIDDEN LANDSCAPES

PHOTOMICROGRAPHS OF AN UNSEEN UNIVERSE

STEVEN BROOKE

I WAS A GRADUATE STUDENT and later a research associate at the University of Miami's Institute for Molecular and Cellular Evolution, under the direction of Dr. Sidney W. Fox, a world leader in the research on chemical evolution and the origin of cells. My research focused on the origins of cellular compartmentalization, models of primordial cellular reproduction, and the origins of cellular communication and behavior. My photomicrographs and scanning electron micrographs of proteinoid microspheres, laboratory models of primordial cells produced under simulated prebiotic conditions, appeared in dozens of journals and textbooks. Below is a scanning electron micrograph of a compartmentalized microsphere produced under such conditions.

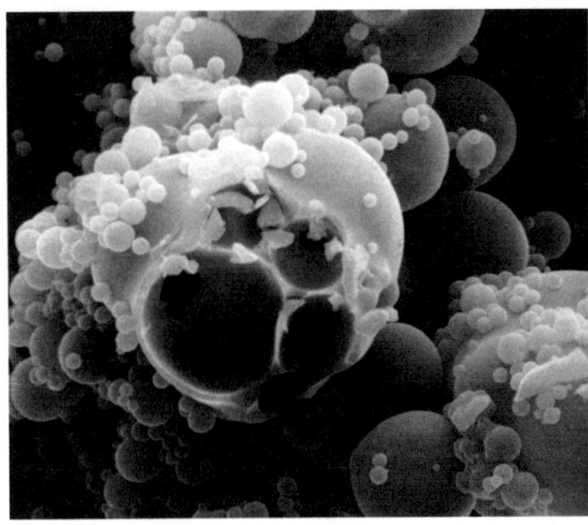

While still at the Institute, I was inspired to produce the photomicrographs in this book after I met Jean Ann Welch, a tropical realist painter of extraordinary vision and masterful technique.

When I asked my friends in the Miami design community to whom I should show the photomicrographs, one name was at the top of everyone's list—Dennis Jenkins. He was considered the most innovative and creative designer in Miami. When I presented my prints to him, he immediately understood the personal synthesis of art and science that I was trying to achieve, and quickly sought to incorporate my pieces into his design work, including several installations for Miami's Southeast Banking Corporation.

Traditionally, art and science were thought to be two incompatible disciplines whose aims, methods, and results had little, if anything, in common. In our own time, artists and scientists now routinely draw upon each other's inspirations and skills. We live in an age uniquely conditioned to finding direct aesthetic enjoyment in the geometric and organic patterns in nature; and collaborations between the disciplines have brought countless unseen wonders into our lives.

Interest in shapes and forms existing in nature was almost nonexistent in Western society before the Renaissance. So called primitive art showed more awareness of existing natural art forms than the art of more "civilized" societies of the past. With Leonardo da Vinci, artistic curiosity was matched by scientific research. His notebooks abound in details concerning the mathematical shapes found in nature. This desire to use the shapes underlying all forms in nature can be traced throughout the history of Western art. Cézanne became increasingly interested and absorbed by his research into these shapes, and his studies led to a new approach to problems of space and volume and their application to painting.

Artists at the onset of the Surrealist period were notoriously intelligent and searching. It is not unreasonable to assume that they had acquainted themselves with the results of a century of microscopic morphology. The popular science books of the late 19th century abounded in these illustrations. The success (even to this day) of Ernst Haeckel's *Art Forms in Nature*, first published in 1892, was certainly one of the first indications that the forms introduced to science by microscopy might have a broader aesthetic appeal. Haeckel, himself, actually argued for "soul" in crystals which would indicate to what degree books of this type in the 19th century were infused with imagination and spirit in addition to their scientific content.

In *Philebus of Plato*, Socrates says, "I will try to speak of the beauty of shapes, and I do not mean, as most people would suppose, the shapes of living things, or their imitation in painting, but I mean straight lines and curves and the shapes made from them.... They are not beautiful for any other reason or purpose, as other things are, but are eternally, and by their very nature, beautiful, and give a pleasure of their own quite free from the itch of desire; and in this way colors can give similar pleasure." The forms which Plato says are of their nature beautiful are clearly geometric forms. This passage is often quoted by supporters of pure abstraction and underscores one aspect of appreciation of the beauties found within the microscopic world.

Kenneth Clark suggests that landscape painting is the furthest from the Platonic ideal of geometric shapes. However, the microscope and telescope have greatly enlarged our range of vision; and that which we can see with our unaided eyes often fails to satisfy our imaginations.

Our everyday landscape and that of the amoeba in its own surroundings have fused. That knowledge left a distinct mark on modern painting and on art in general. The work of artists such as Joan Miró, Salvador Dali, Jean Arp, Yves Tanguy, and especially Paul Klee bear this mark. Many of Klee's watercolors literally teem with cellular-type inhabitants, strongly suggesting Klee's familiarity with the microscopic world.

The histories of modernist art and the microscopic field are also entangled in the belief that certain forms had attached to them a privileged status. Both, of course, inherit this belief from an idealist philosophy.

Crystallography provides the unqualified inorganic prototype. Of nature's endlessly beautiful and bizarre forms, perhaps few are more provocative and mysterious than the colorful, flowing world of crystals. Crystals are homogeneous substances bounded by flat surfaces in various geometric shapes. The configurations are determined by the directional forces within the crystals and are related to their atomic structure which is unique and stable for each class of crystal.

The technique for extracting the treasures from the world beneath our threshold of vision is called photomicrography. It involves a pairing of both microscopic and photographic techniques.

The photomicrographs shown here are largely of crystals, chosen not for their particular scientific importance, but solely for their provocative beauty. Many are, in fact, of the same chemical species, only photographed under various lighting conditions and filtration techniques.

As with any photographic experience, specific regions of these boundless microscopic vistas were selected in response to innermost feelings, needs,

memories — mental processes both conscious and subconscious. My long time fascination with the landscapes and architecture of science fiction book cover art was always informative. In this universe, with no apparent scale, I injected a sense of human dimension, often with a horizon line to function as a visual and psychological anchor.

Through the objective lens of the microscope, the eye hovers over crystalline cities and seemingly ancient ruins; it breaks through lattice-like barriers that open onto infinite space; it glides along translucent shards of crystals, glowing from within; and it navigates through galactic formations inhabited by forms that seem to exist on the border between the inanimate and animate worlds.

Leonardo wrote that looking at amorphous patterns (of plaster walls, for example) and allowing the mind to play upon them, inventing one subject after another, helped to stimulate imaginative seeing. Nature must be frequently contemplated anew if we are to continually recharge our creative forces and sense of astonishment. Some mysterious operation takes place within us when we perceive these natural syllables which have the ability to penetrate deep into our unexplored hearts and reveal the correspondence of all structures.

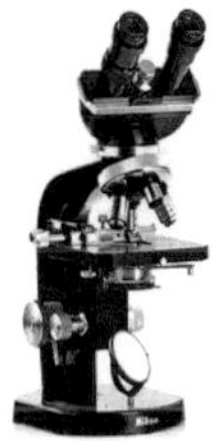

T ECHNICAL NOTES

The photomicrographs were produced with either a Leica 35mm camera or a Sinar 4x5 view camera mounted on a Zeiss polarizing microscope with a variety of objectives, polarizing filters and light sources. Films included 35mm Kodachrome and 4x5 Ektachrome. Magnifications range from 10x to 600x. Materials used include amino acids such as alanine and leucine, potassium ferrocyanide, picric acid, tris buffer, manganese and calcium salts, B vitamins, dyes, and other organic and inorganic salts. Various means of purifying the chemicals were used to enhance their optical properties.

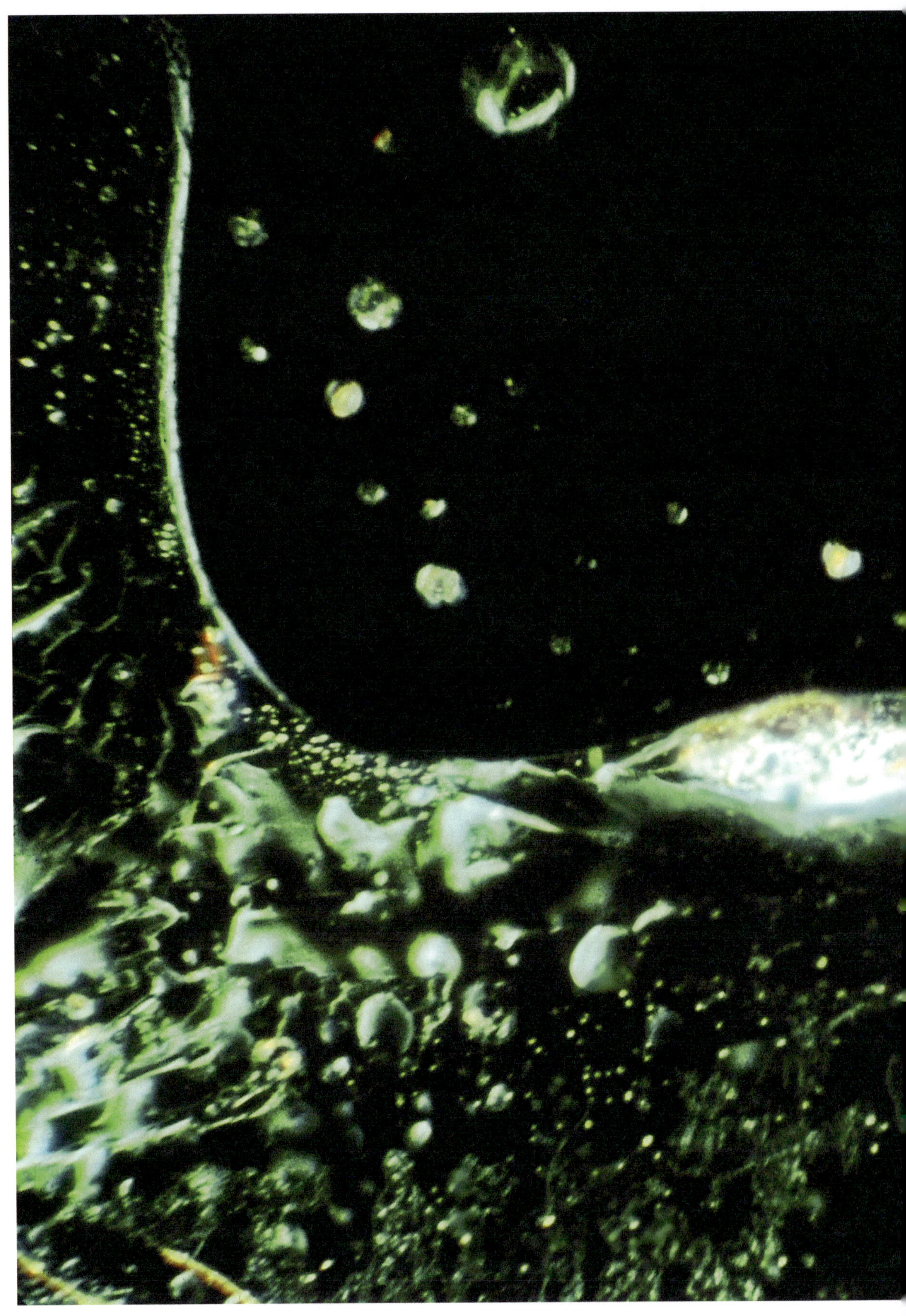

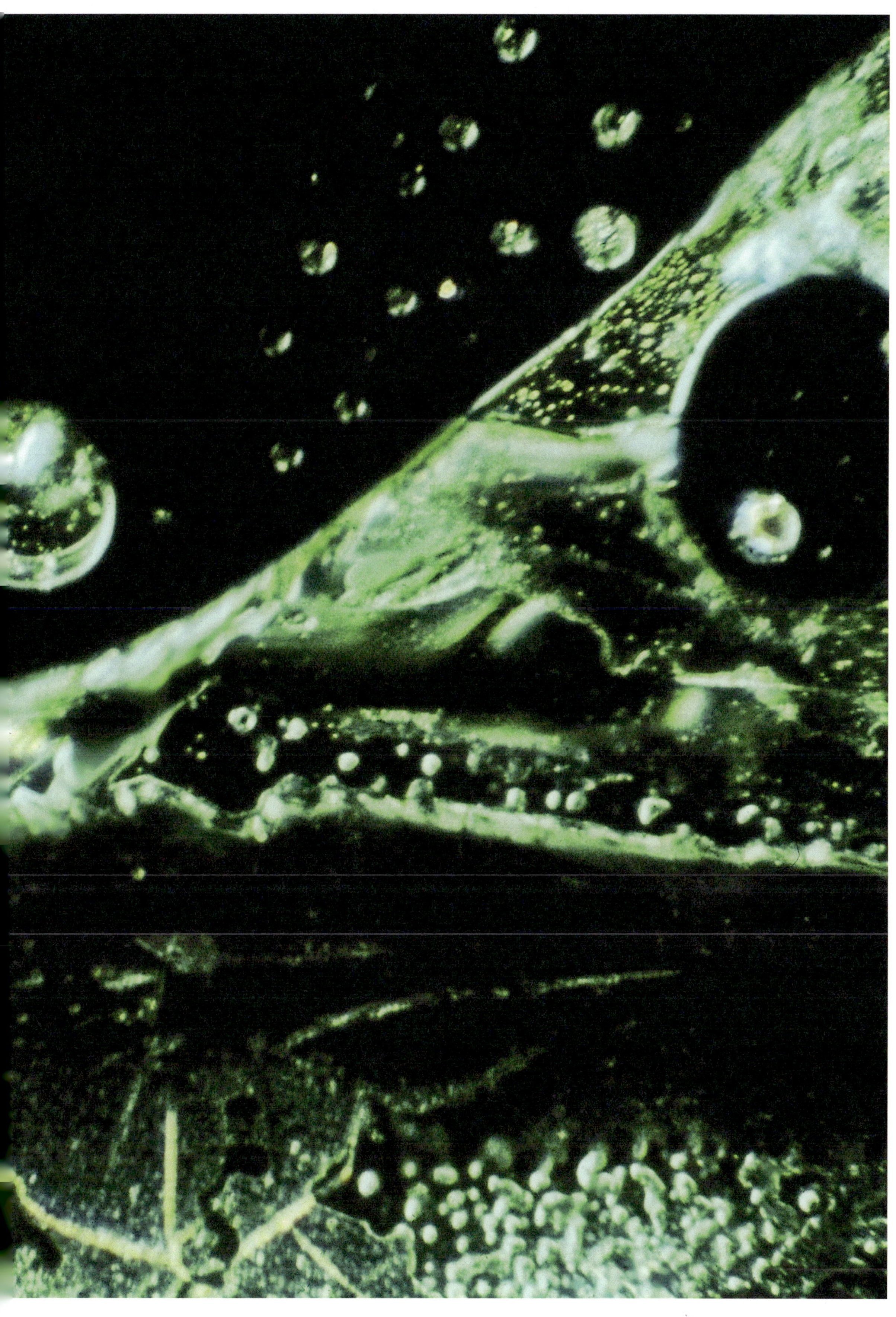

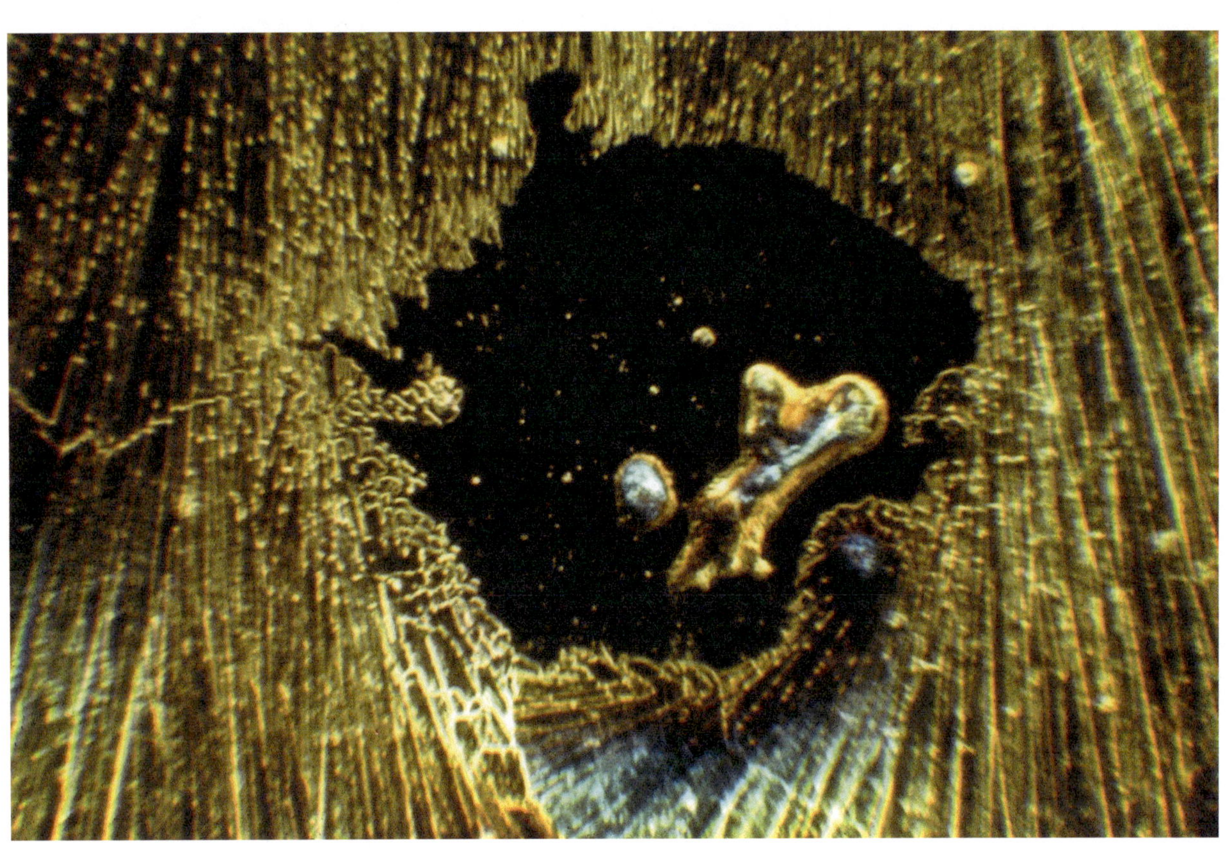

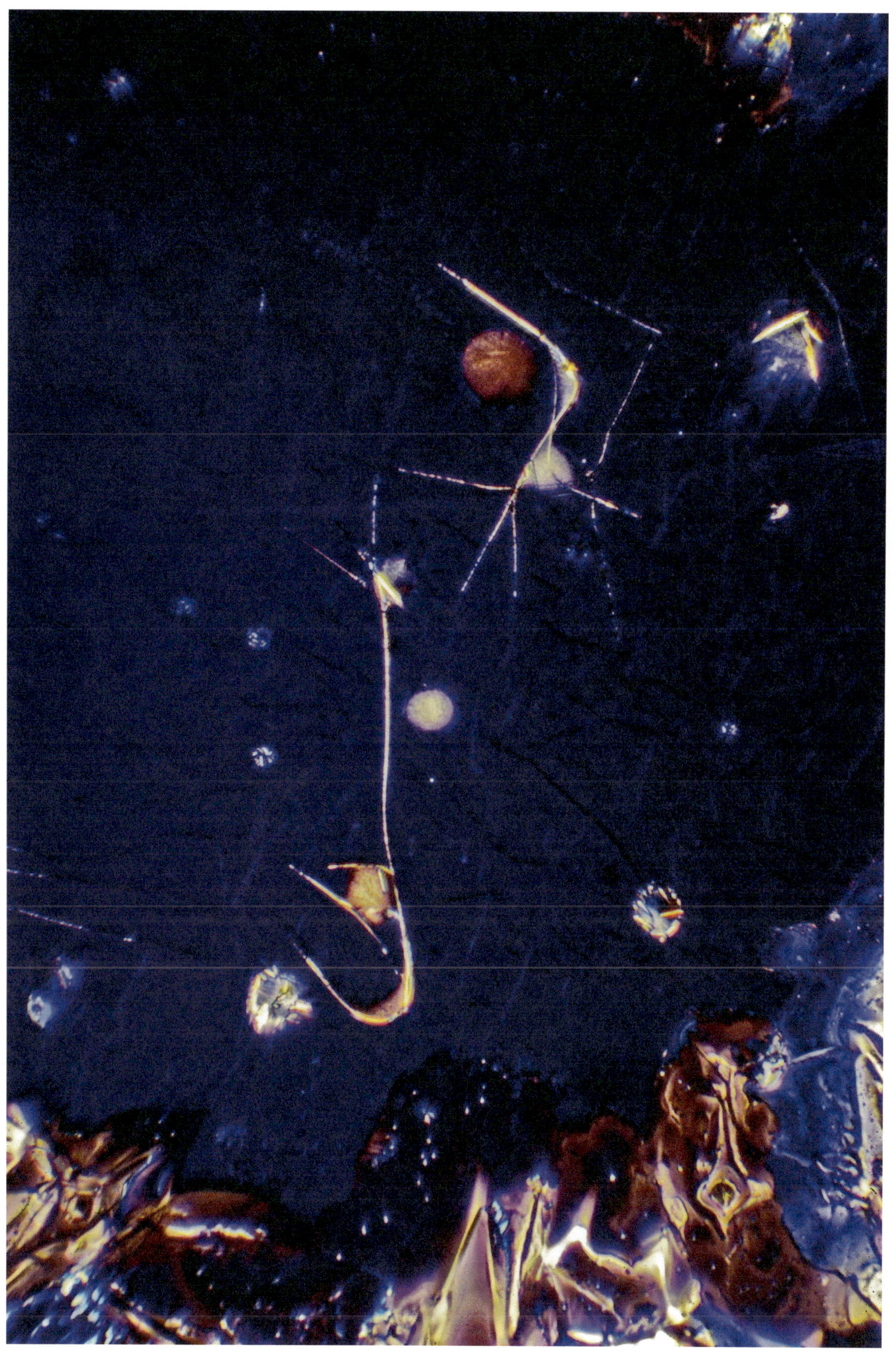

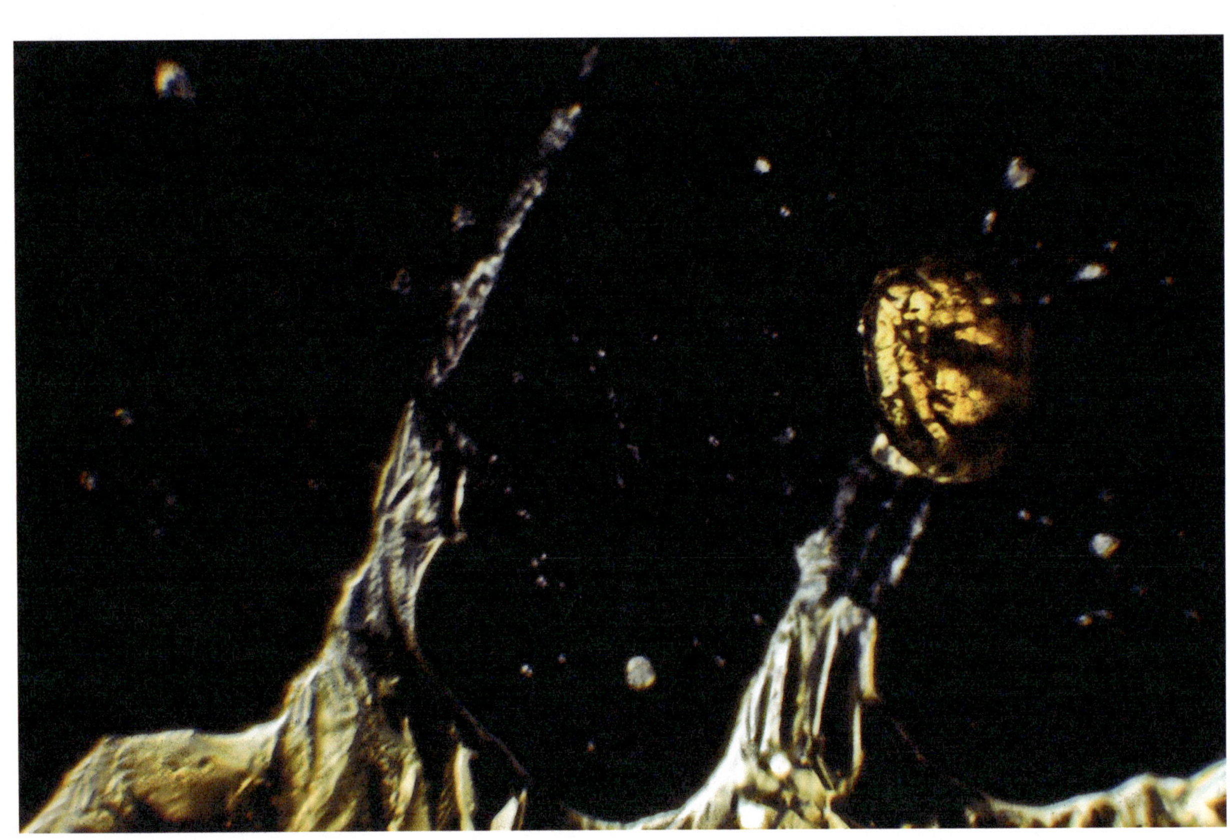

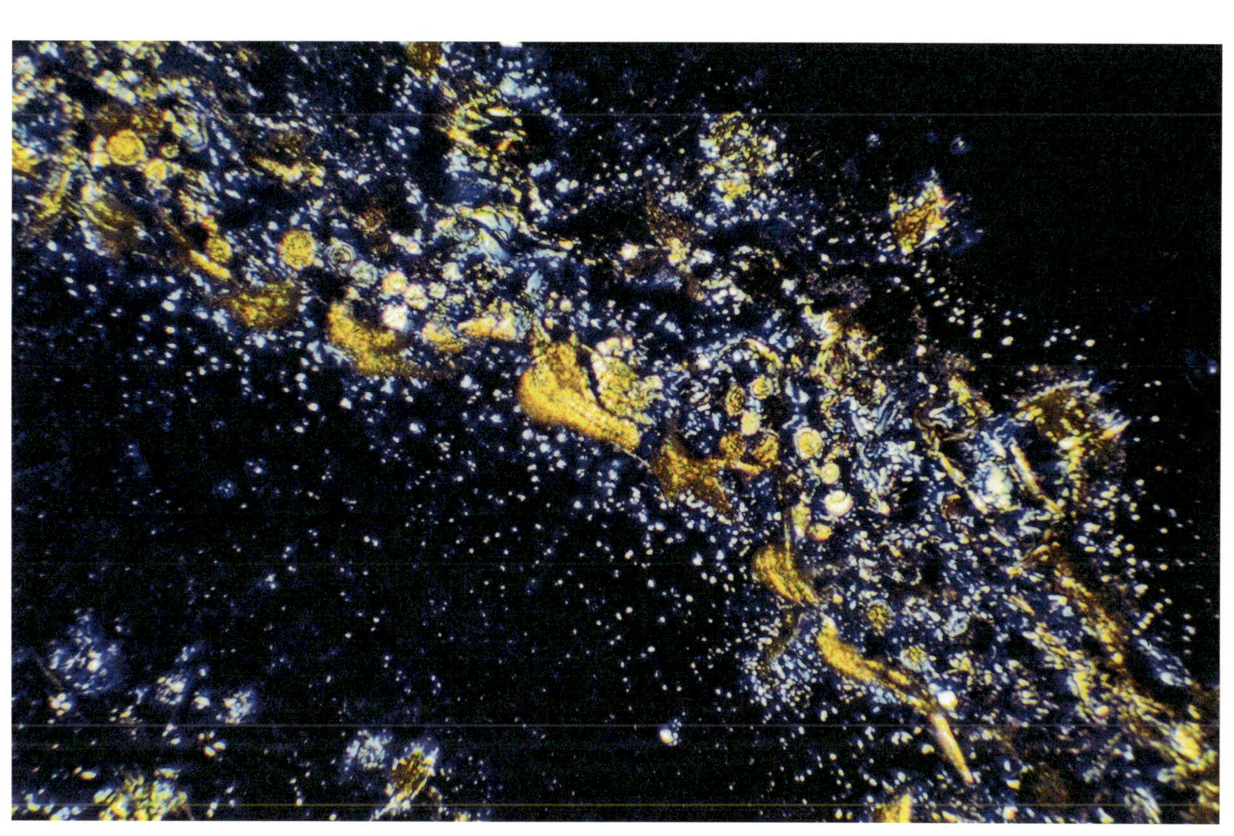

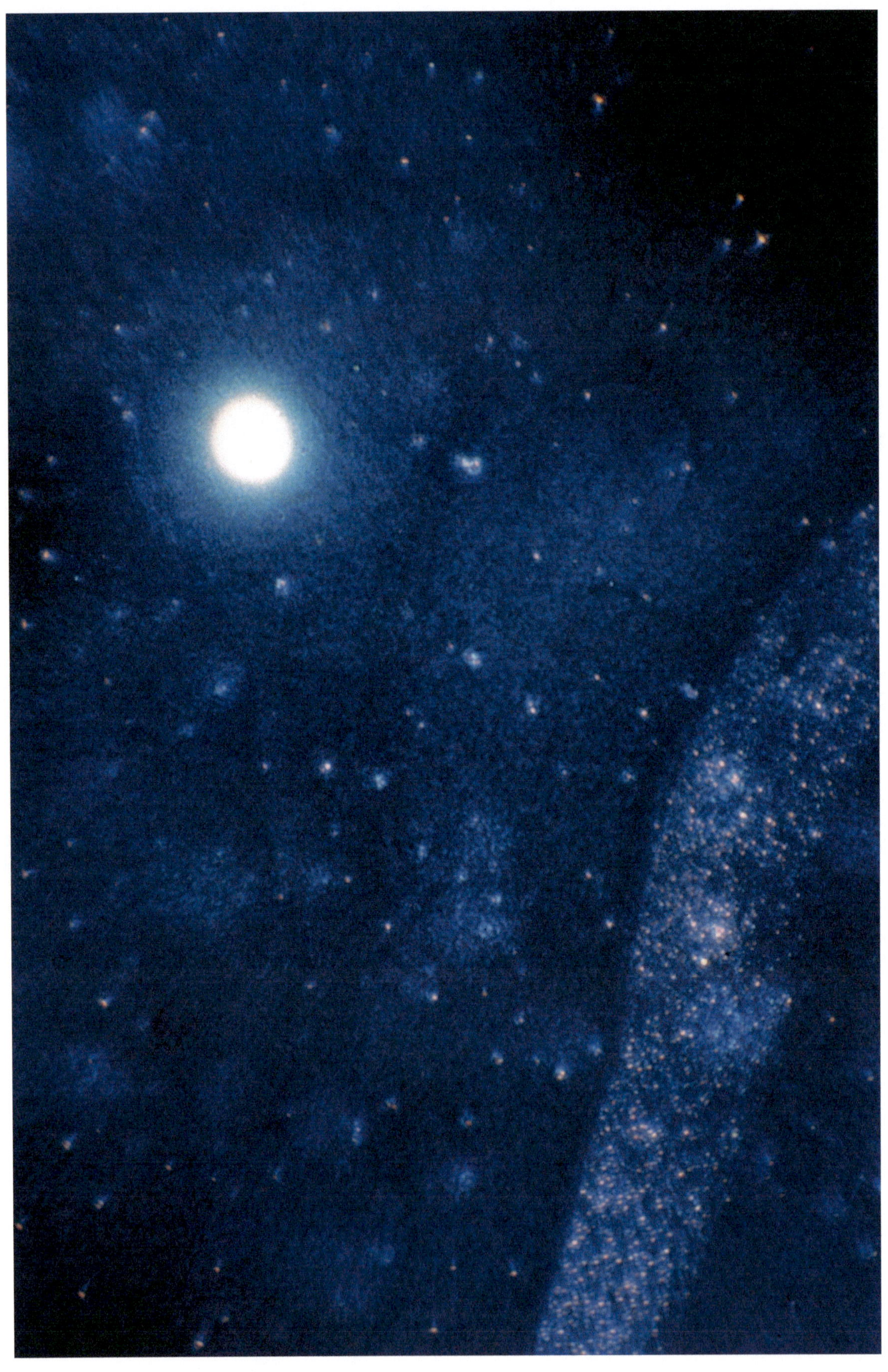

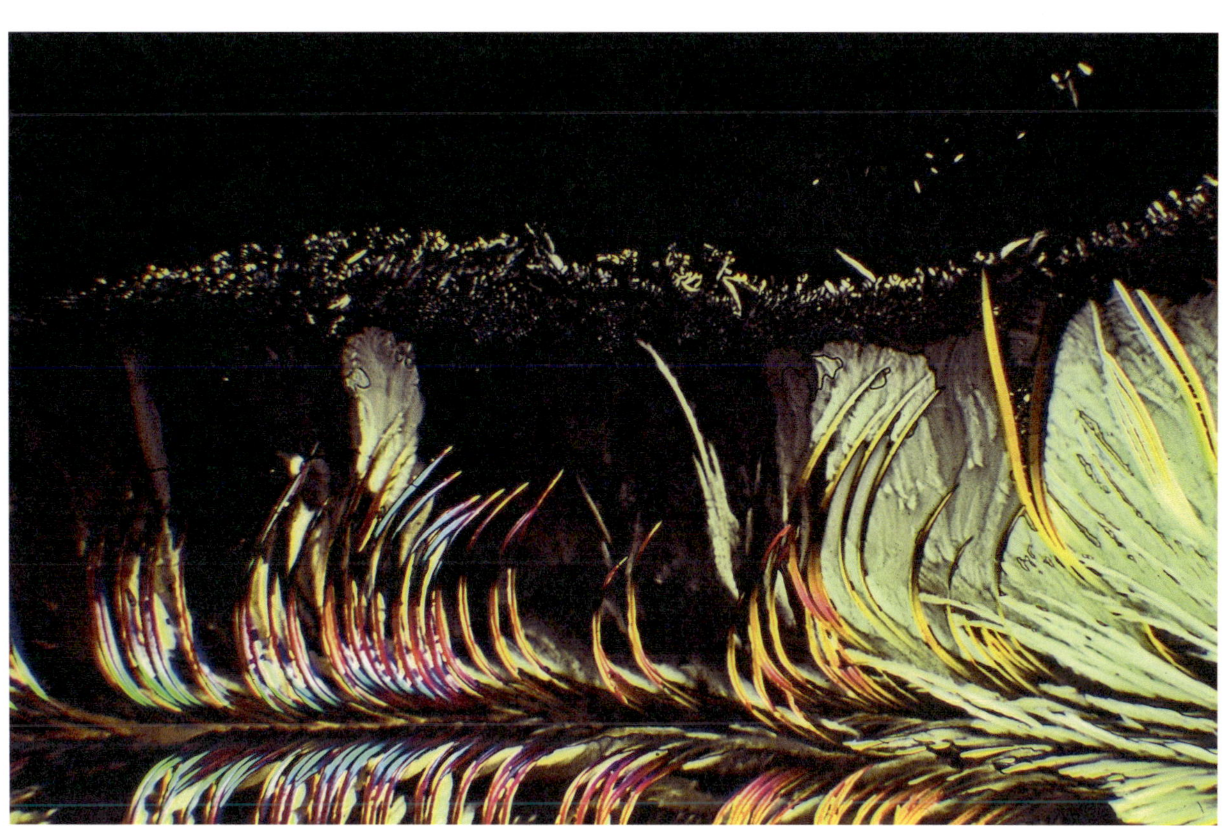

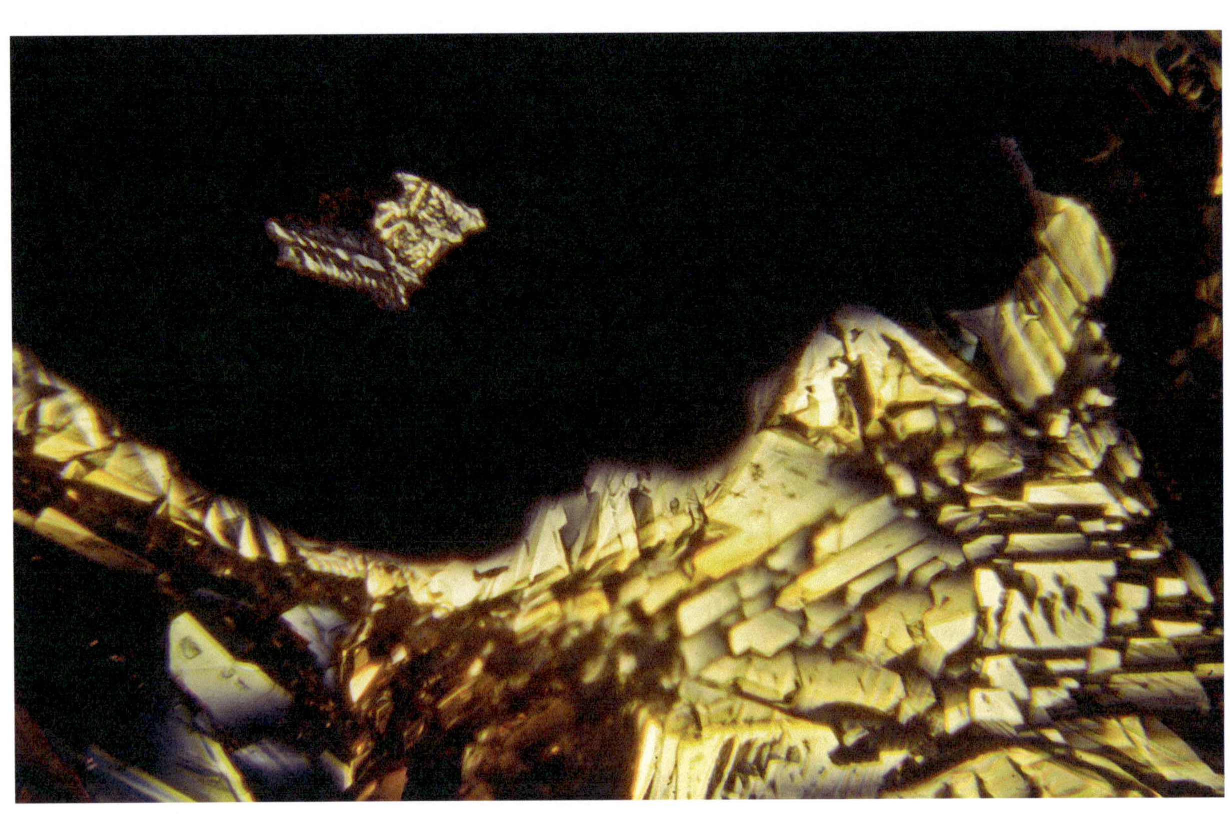

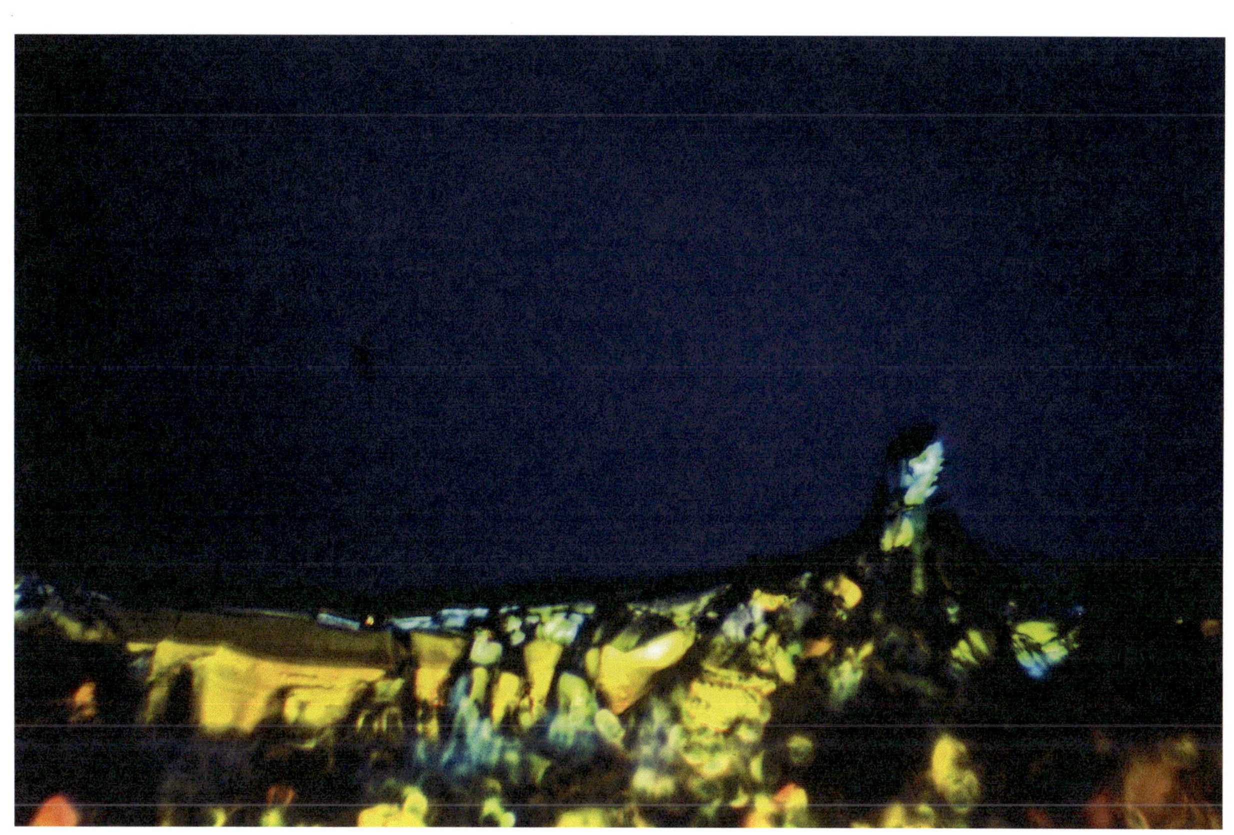

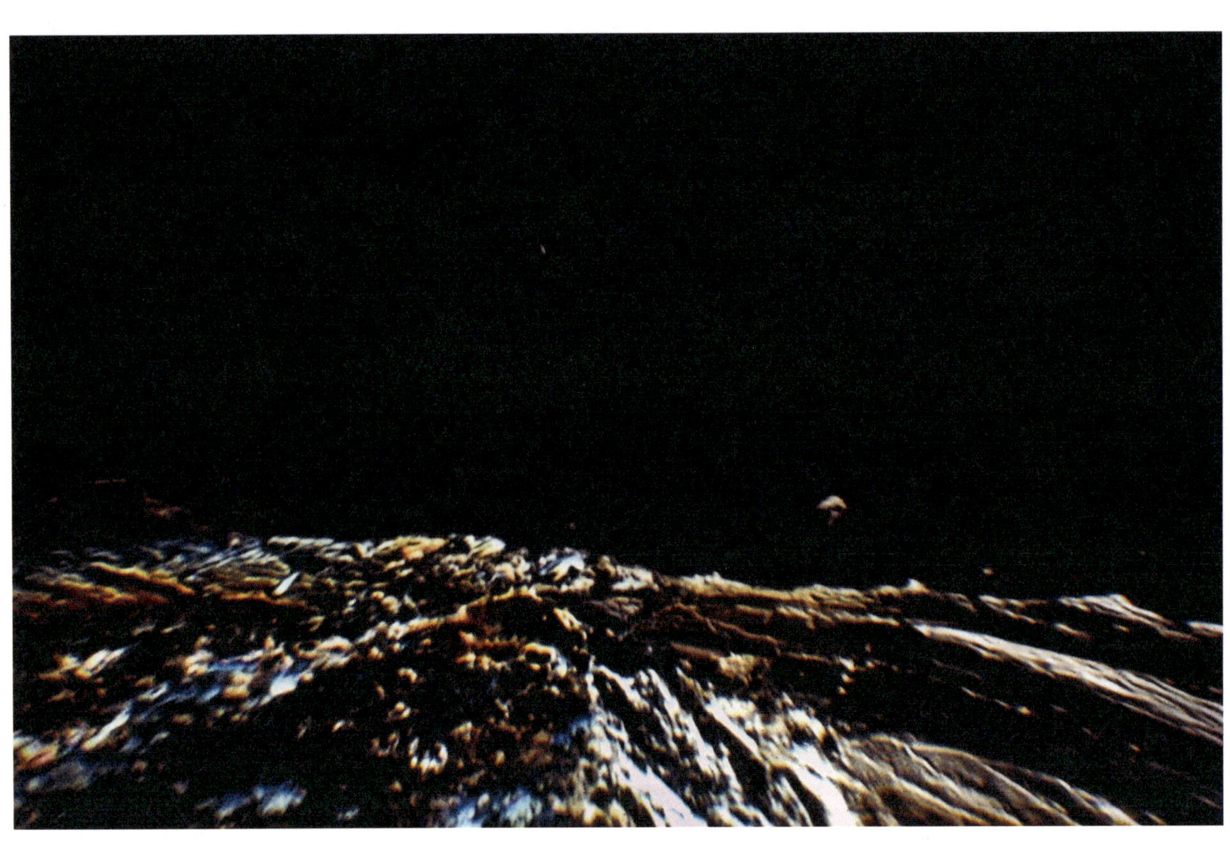

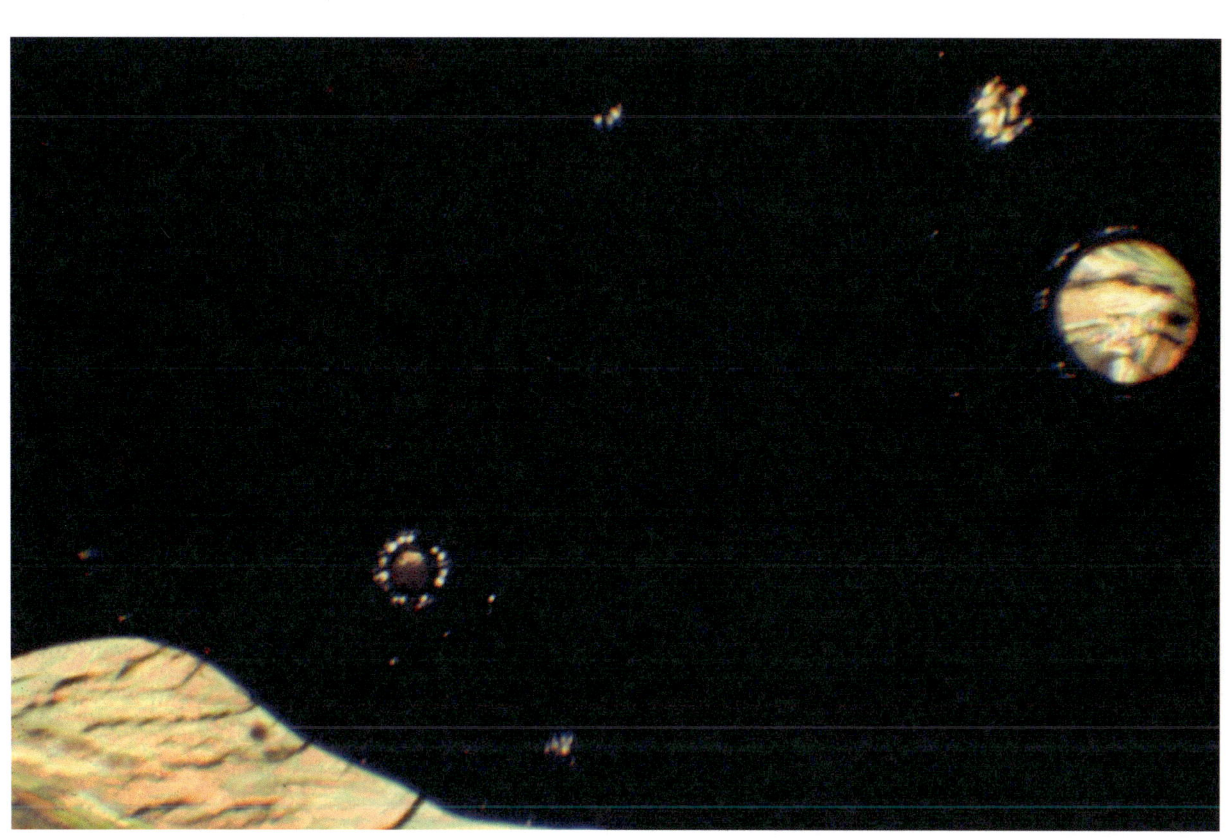

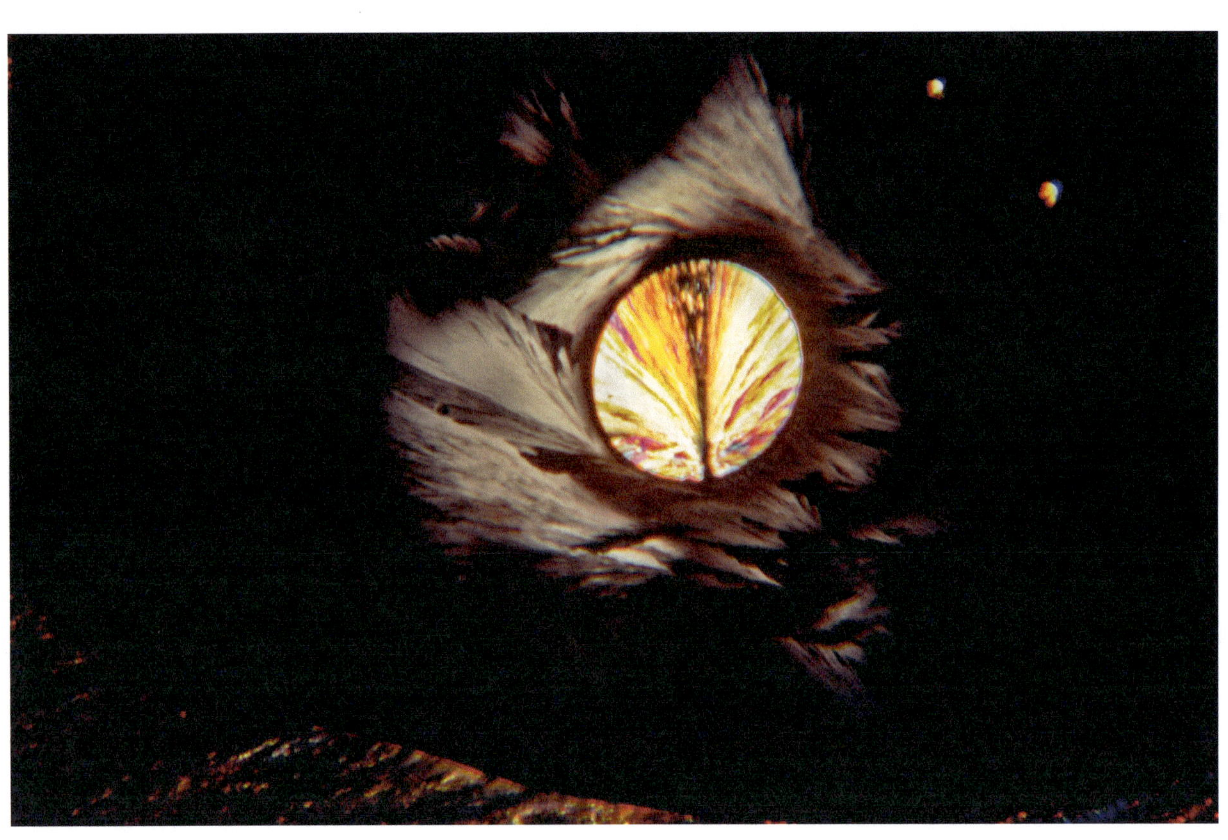

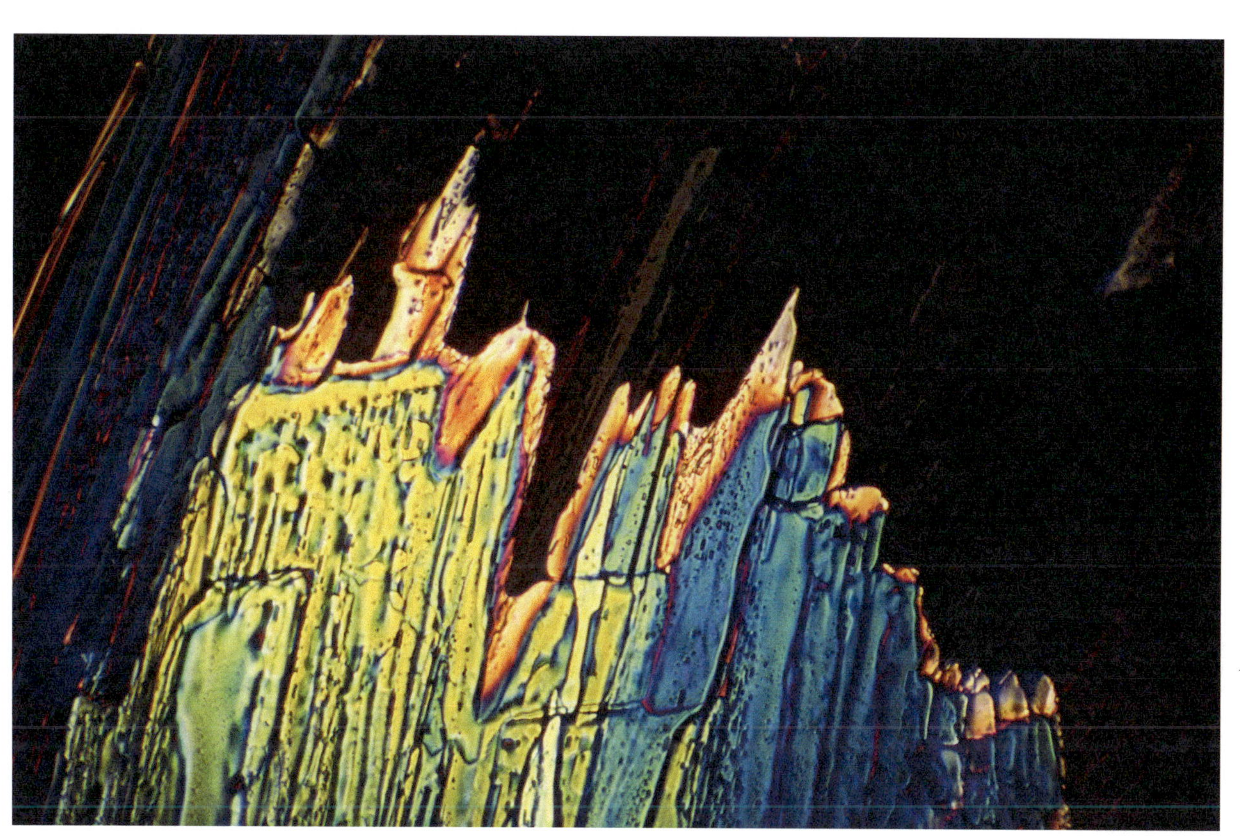

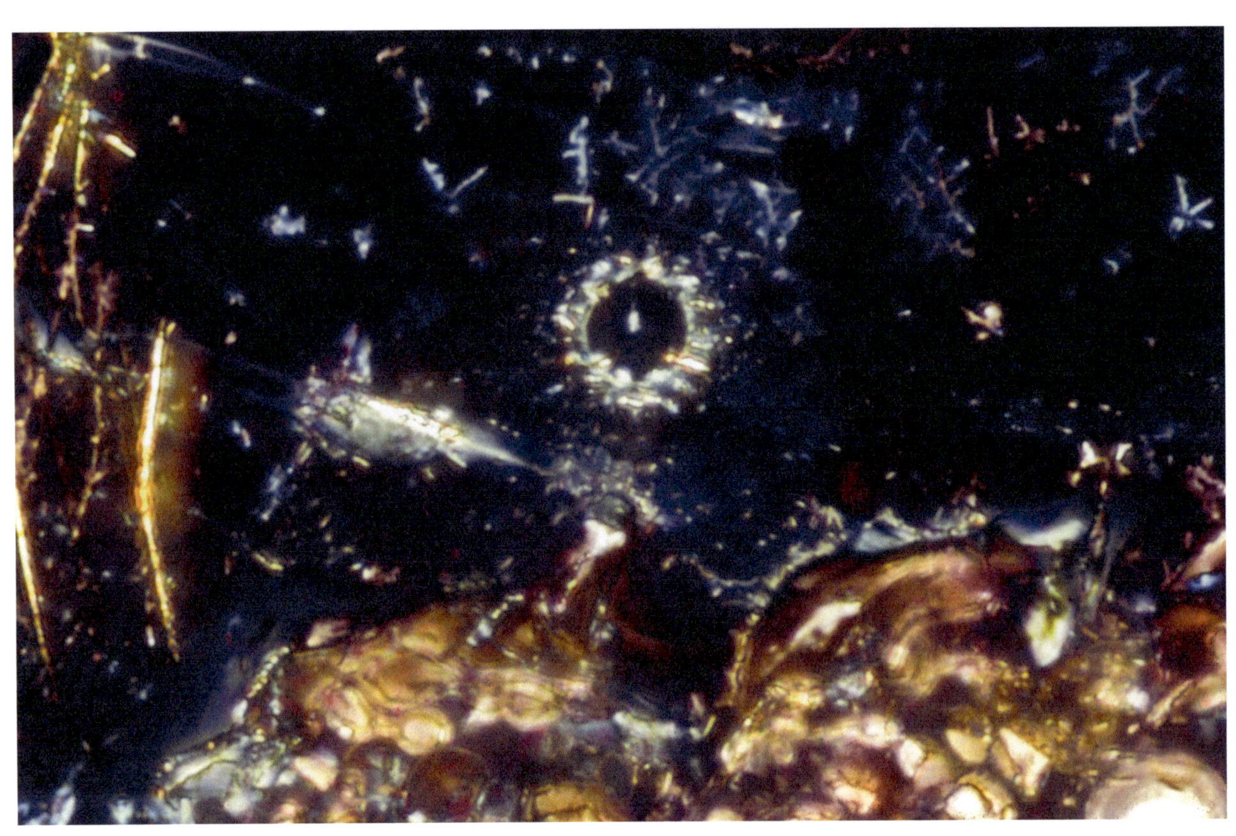

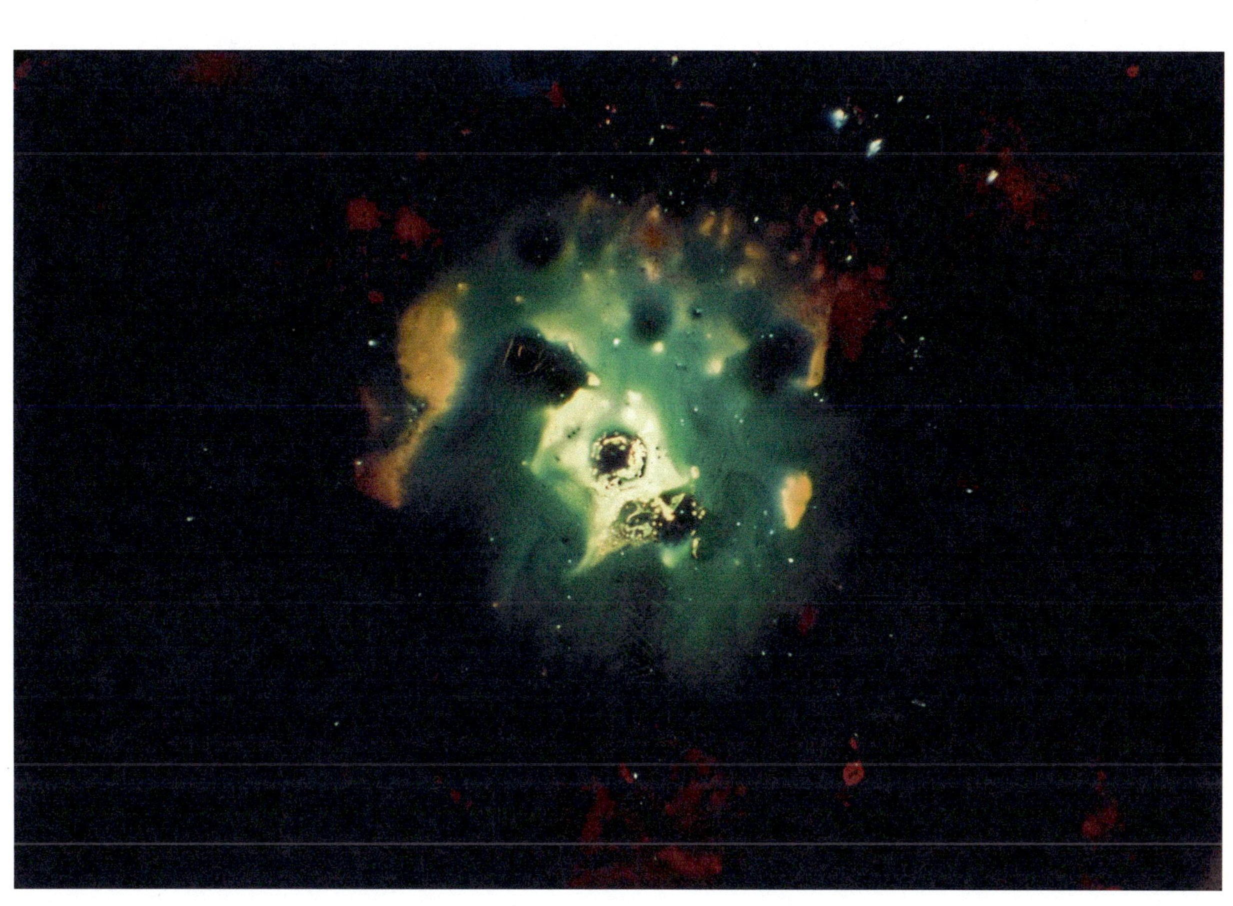

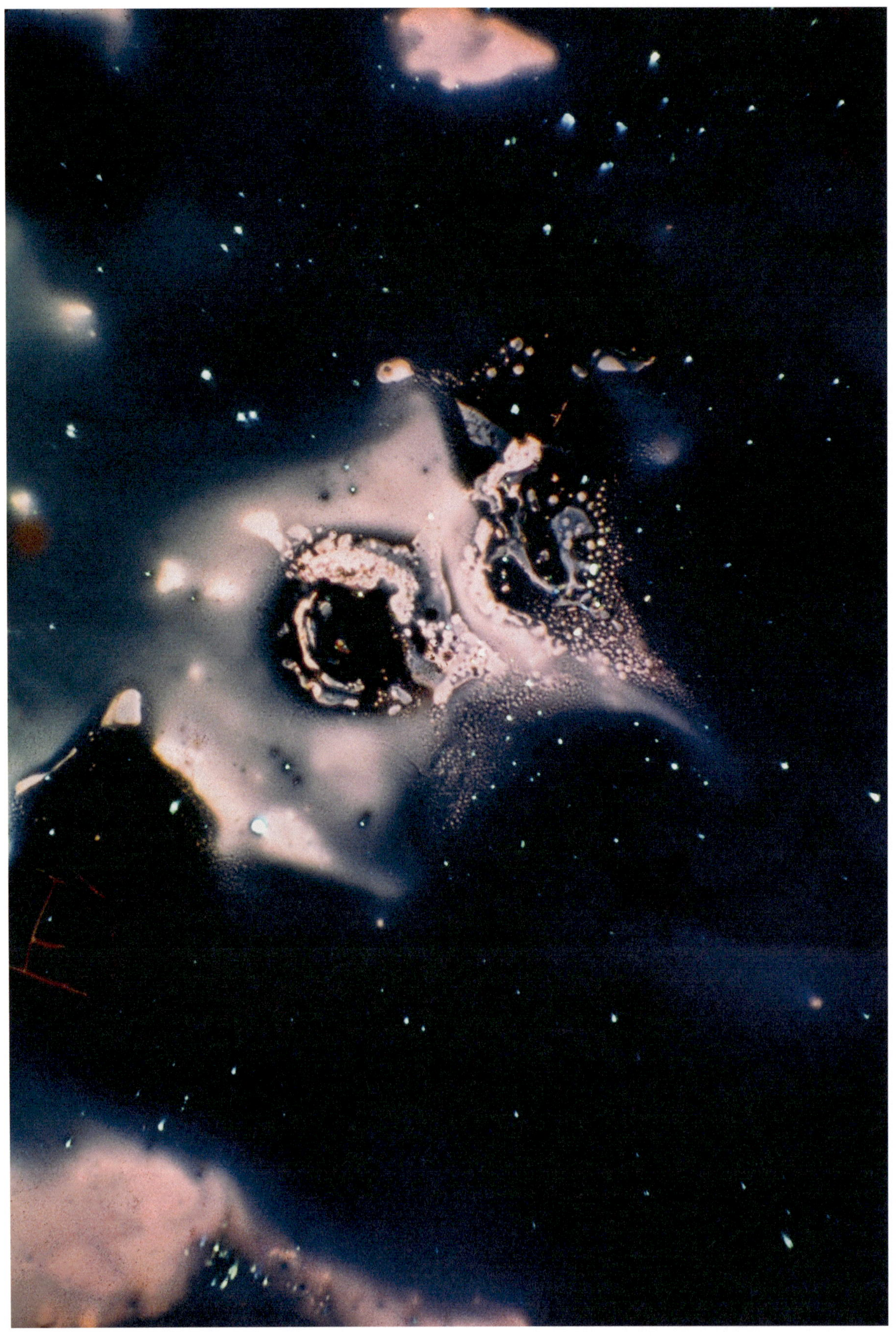

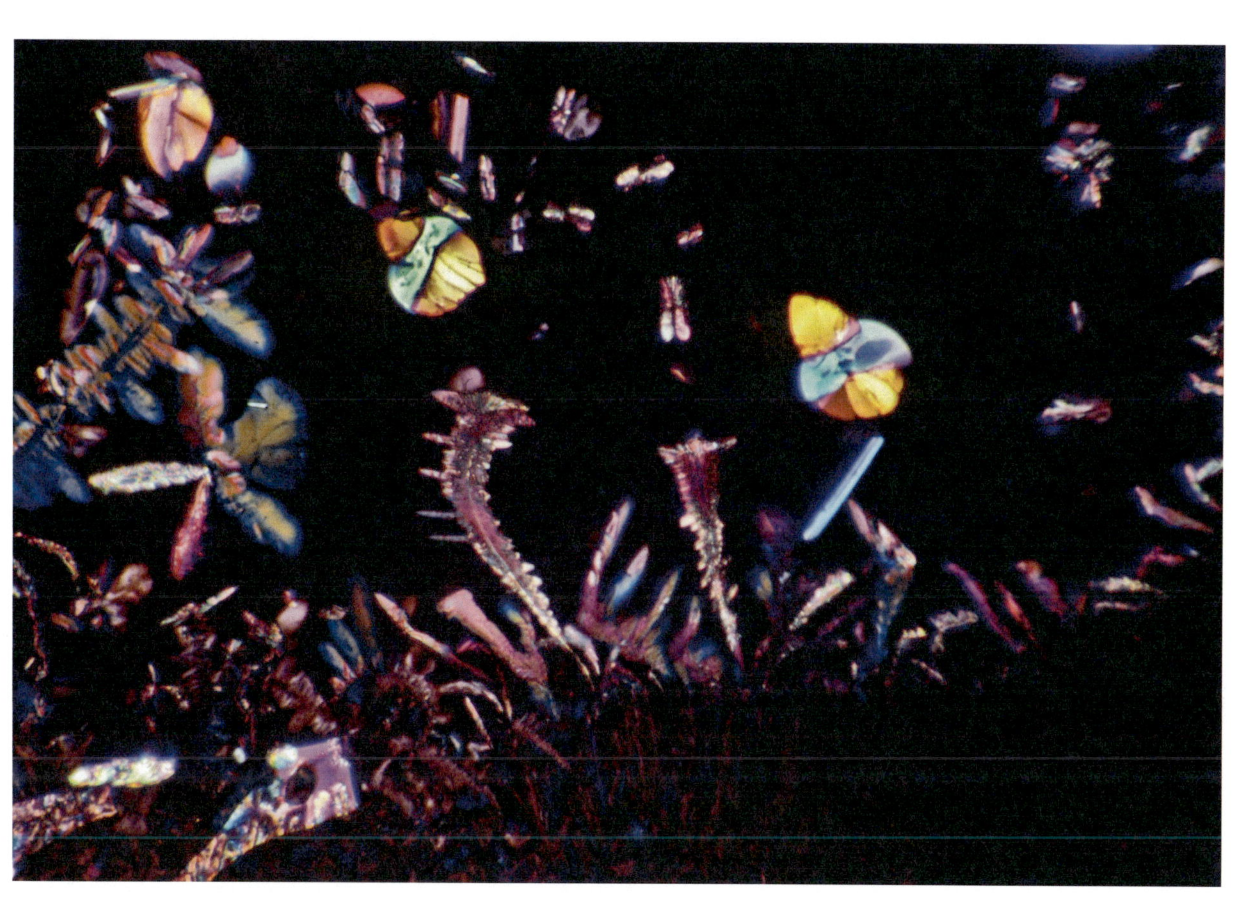

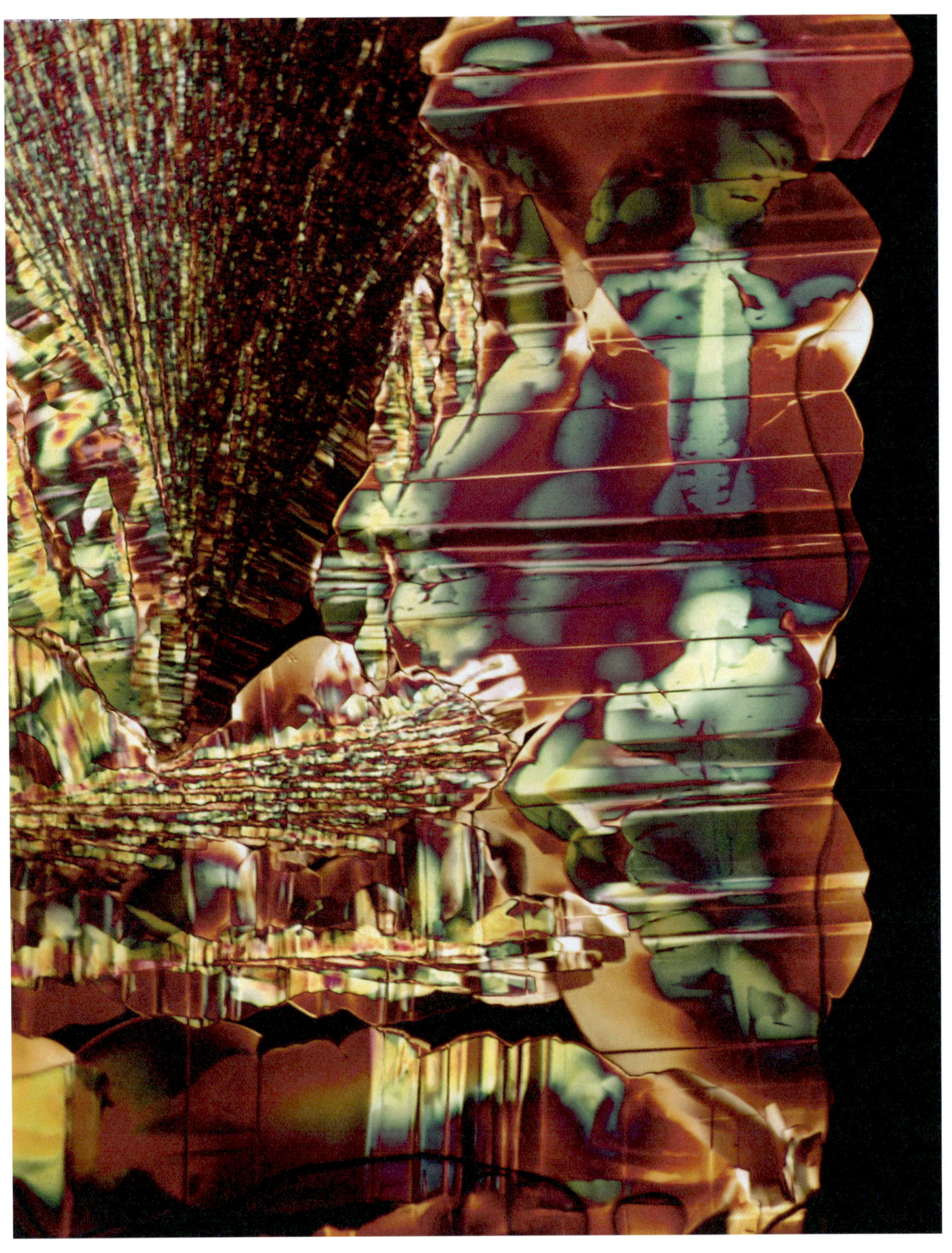

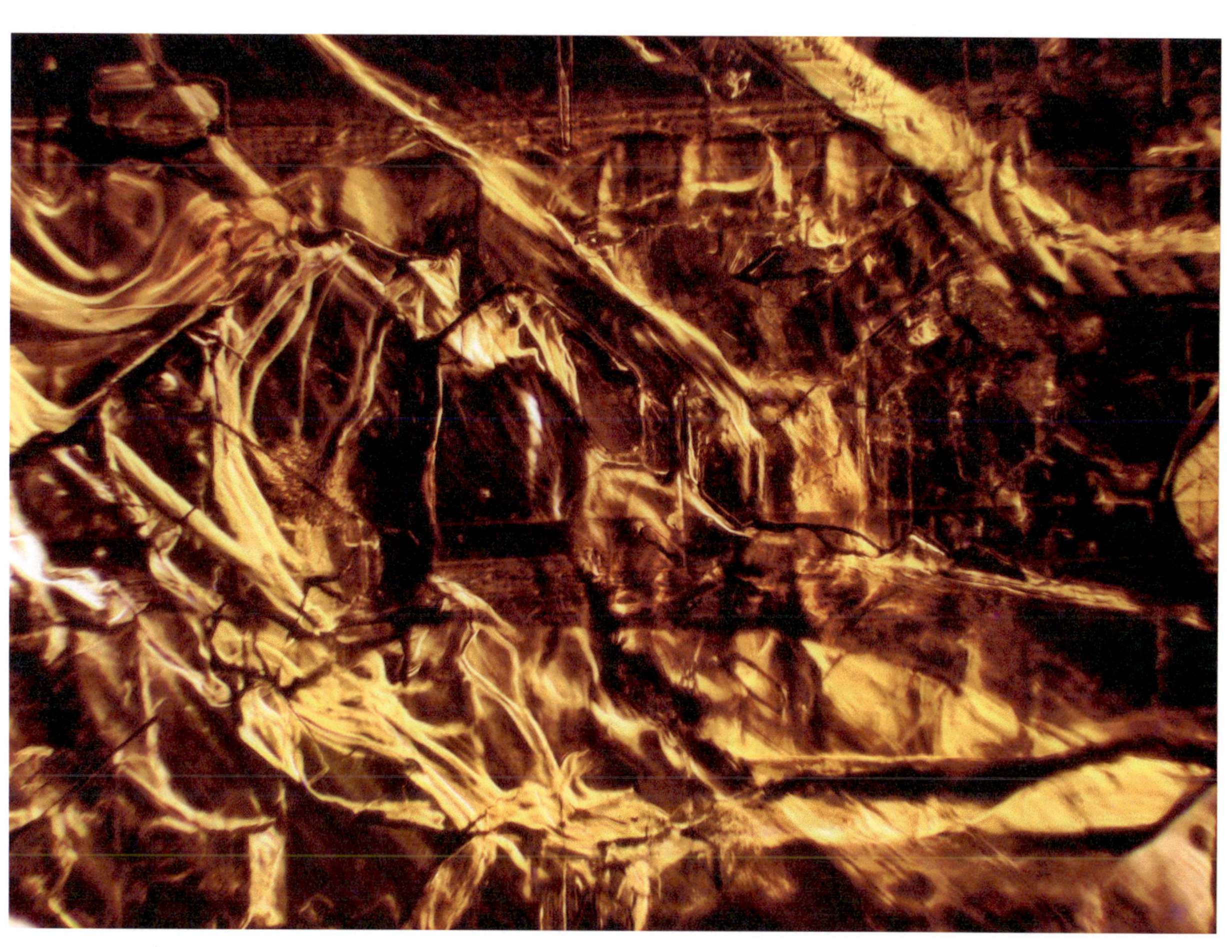

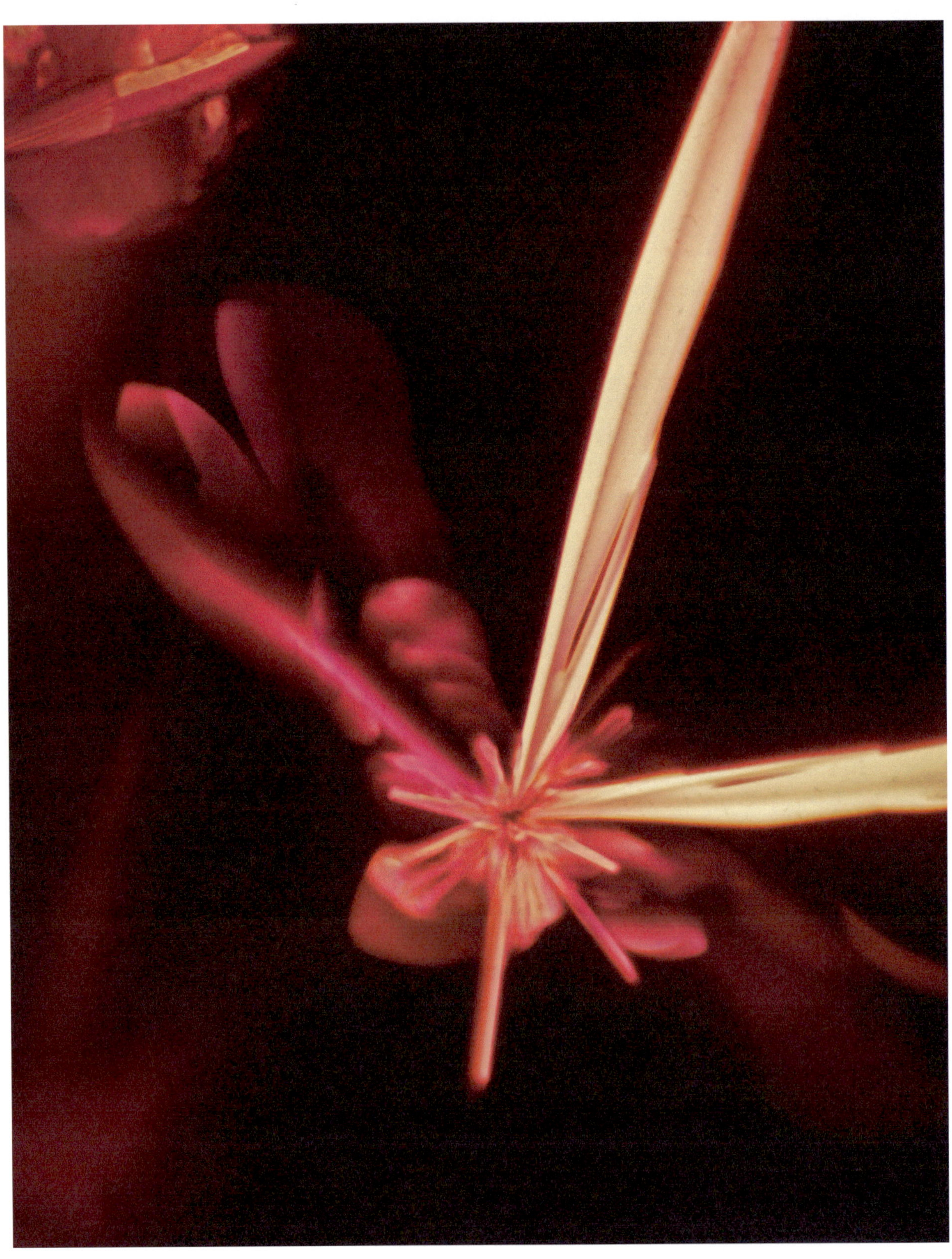

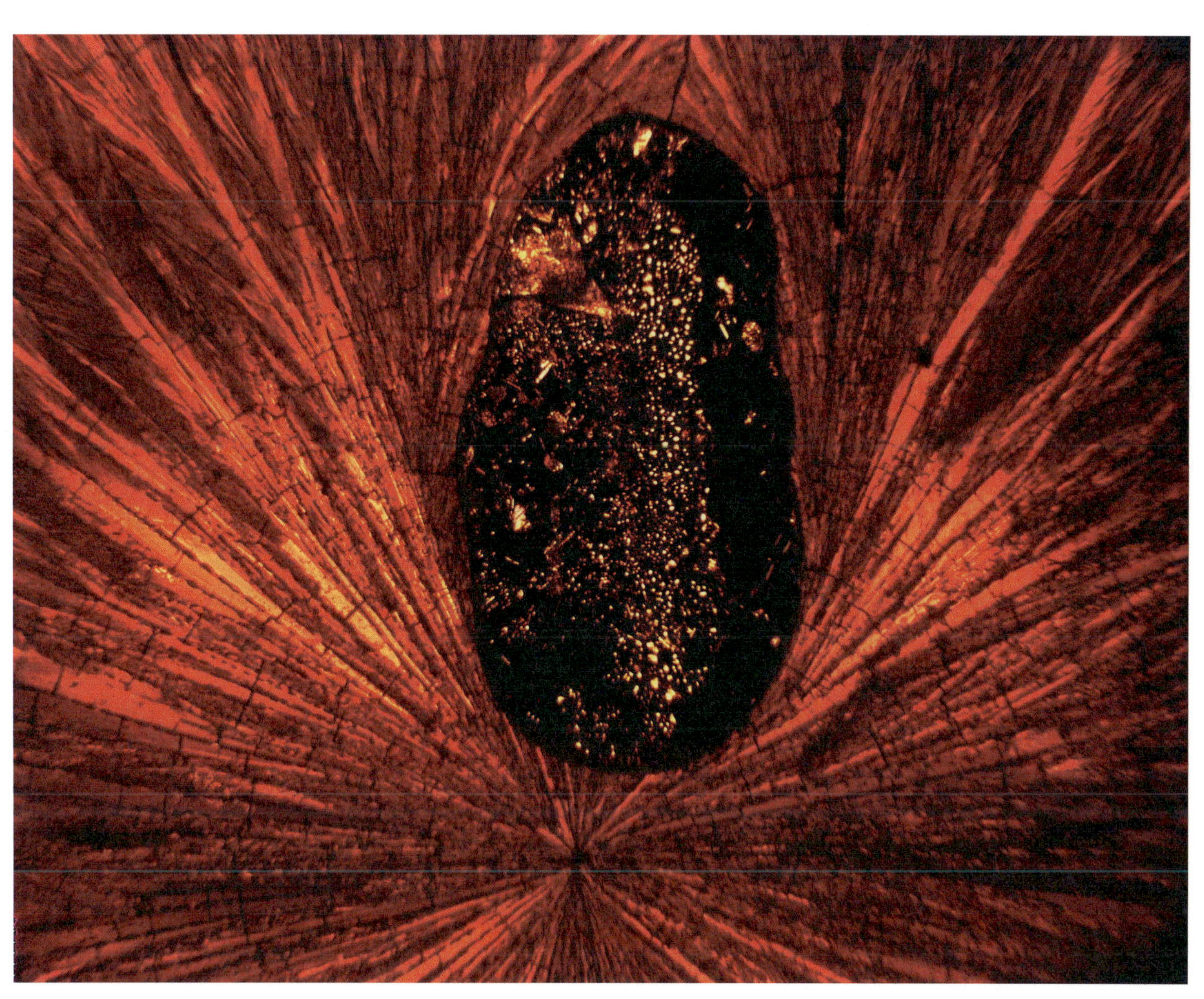

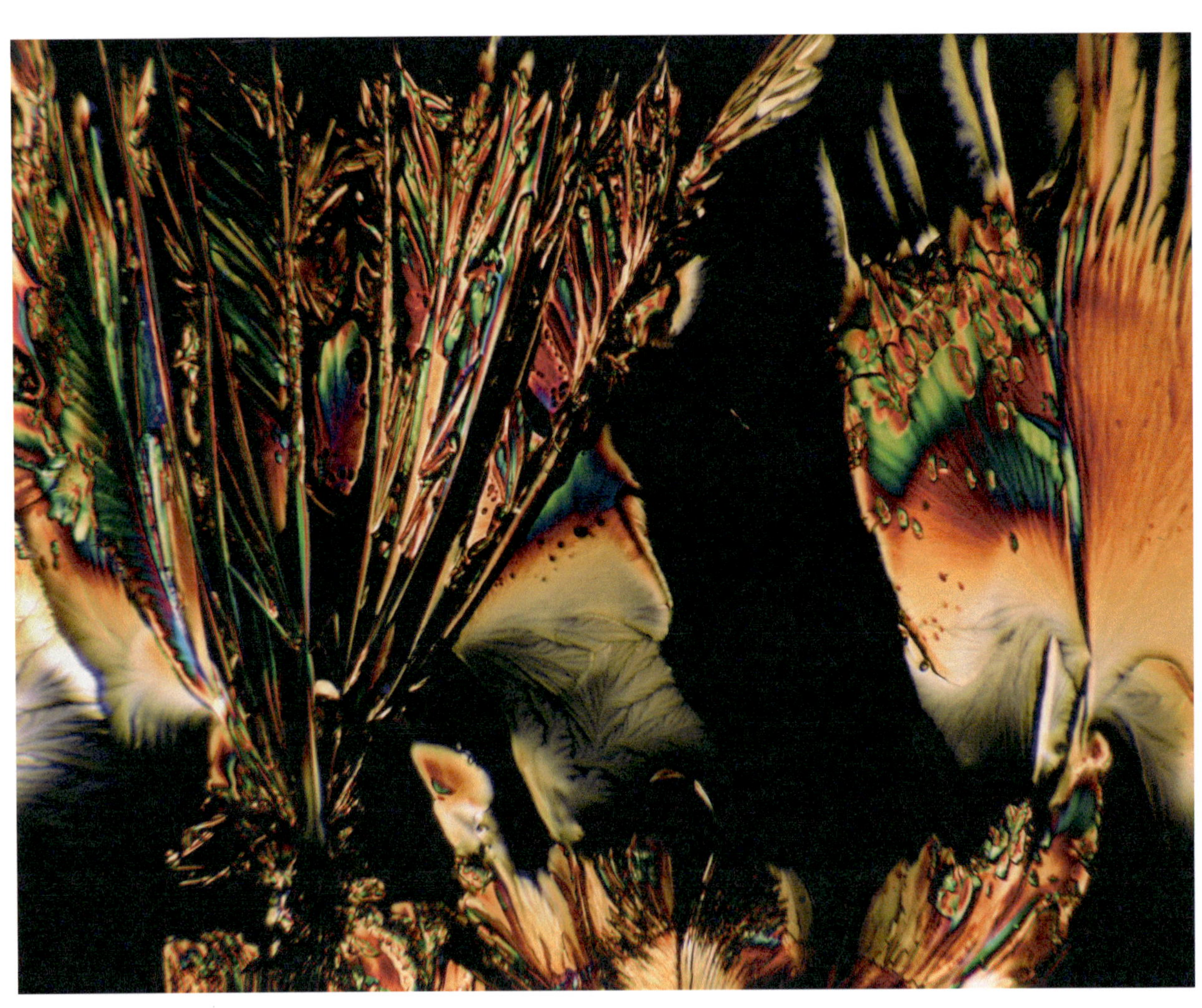

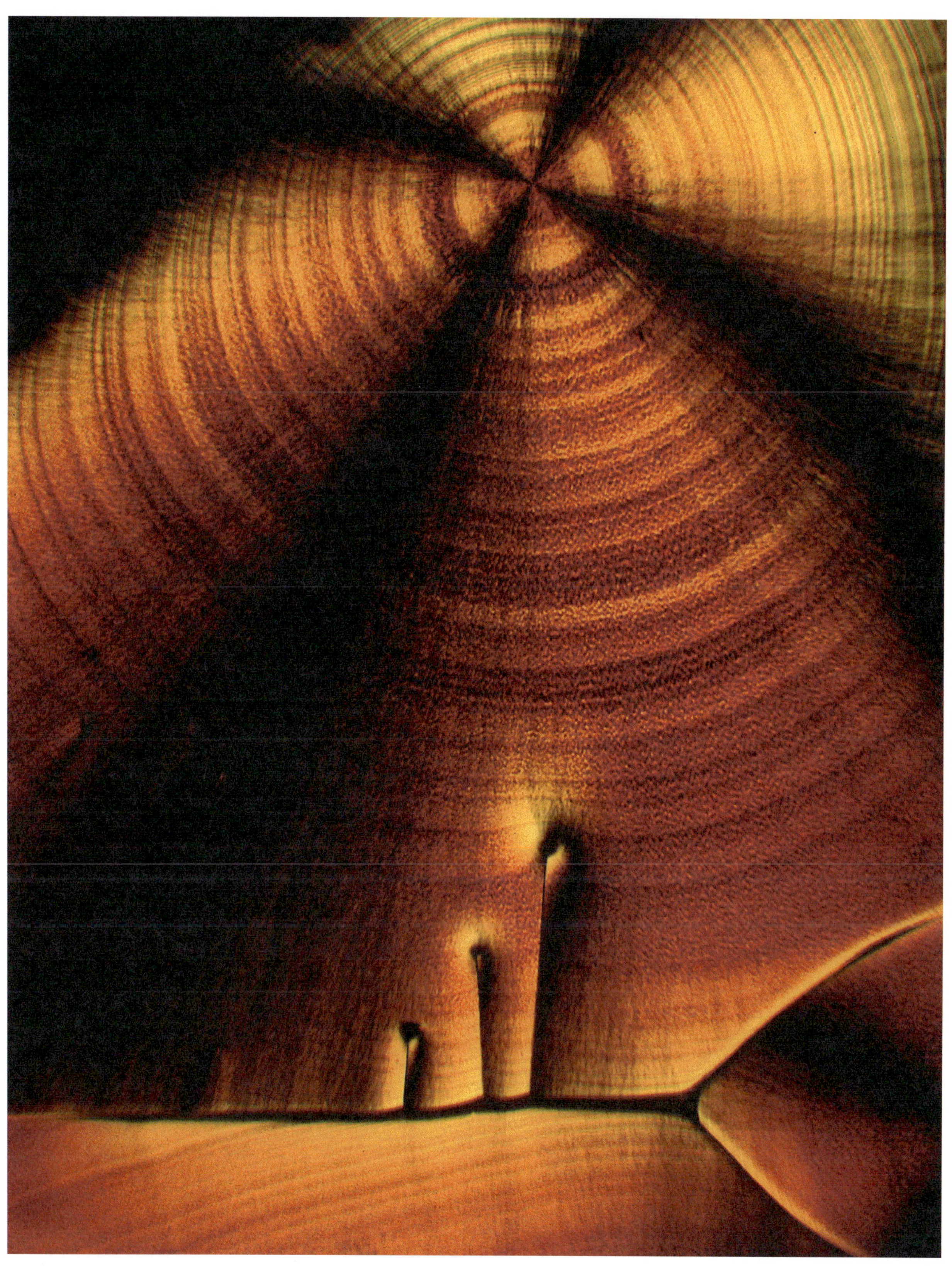

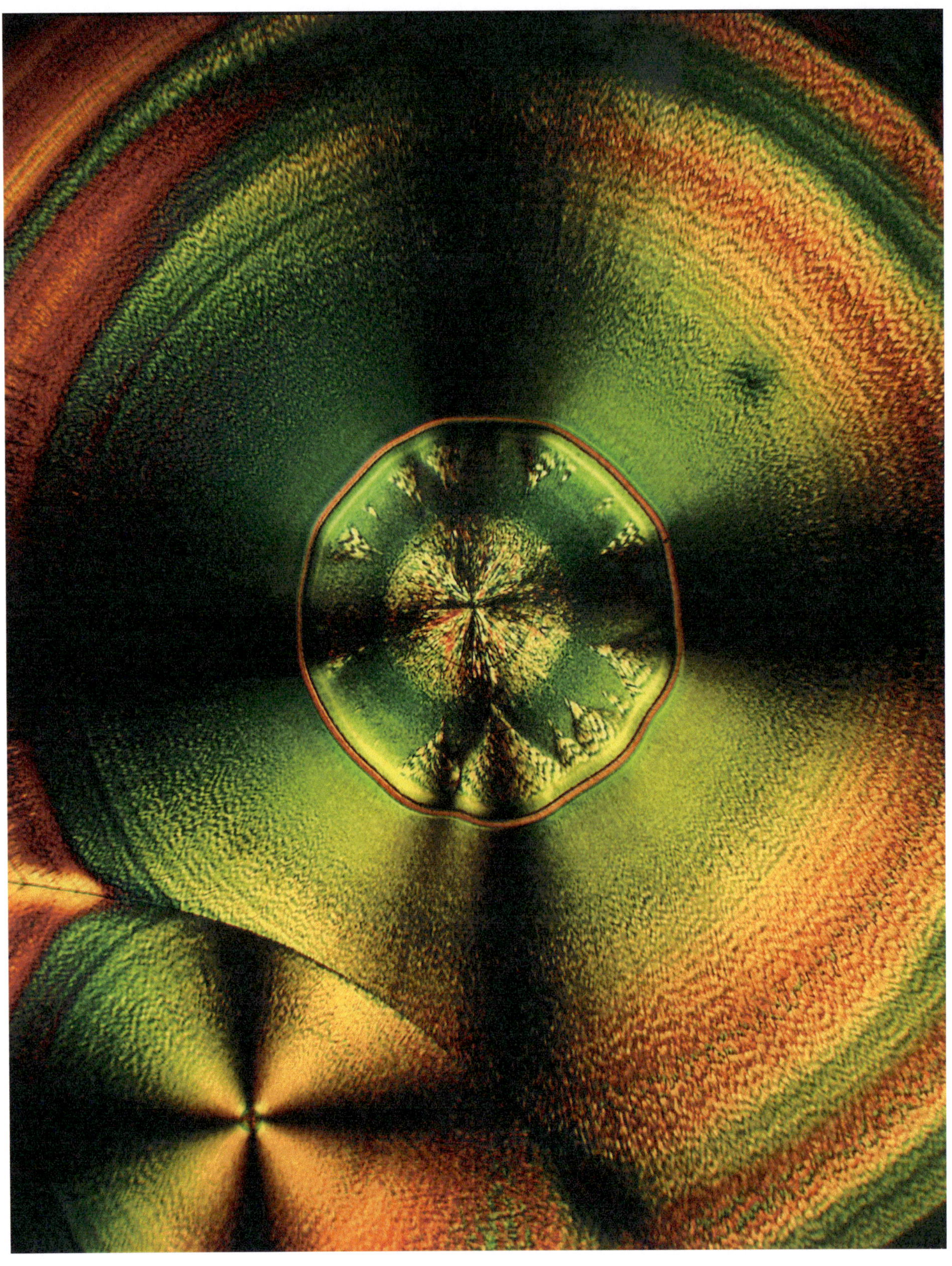

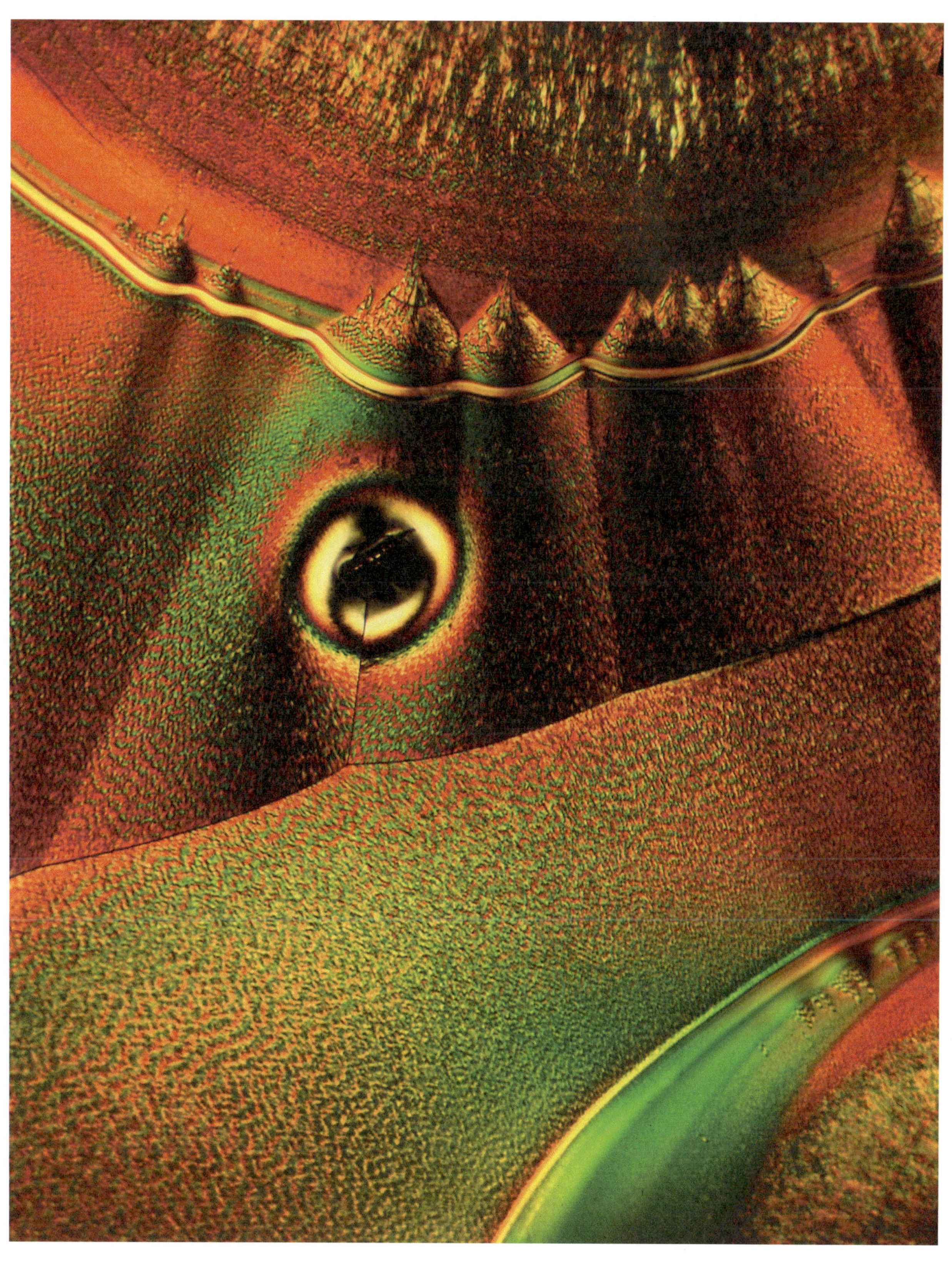

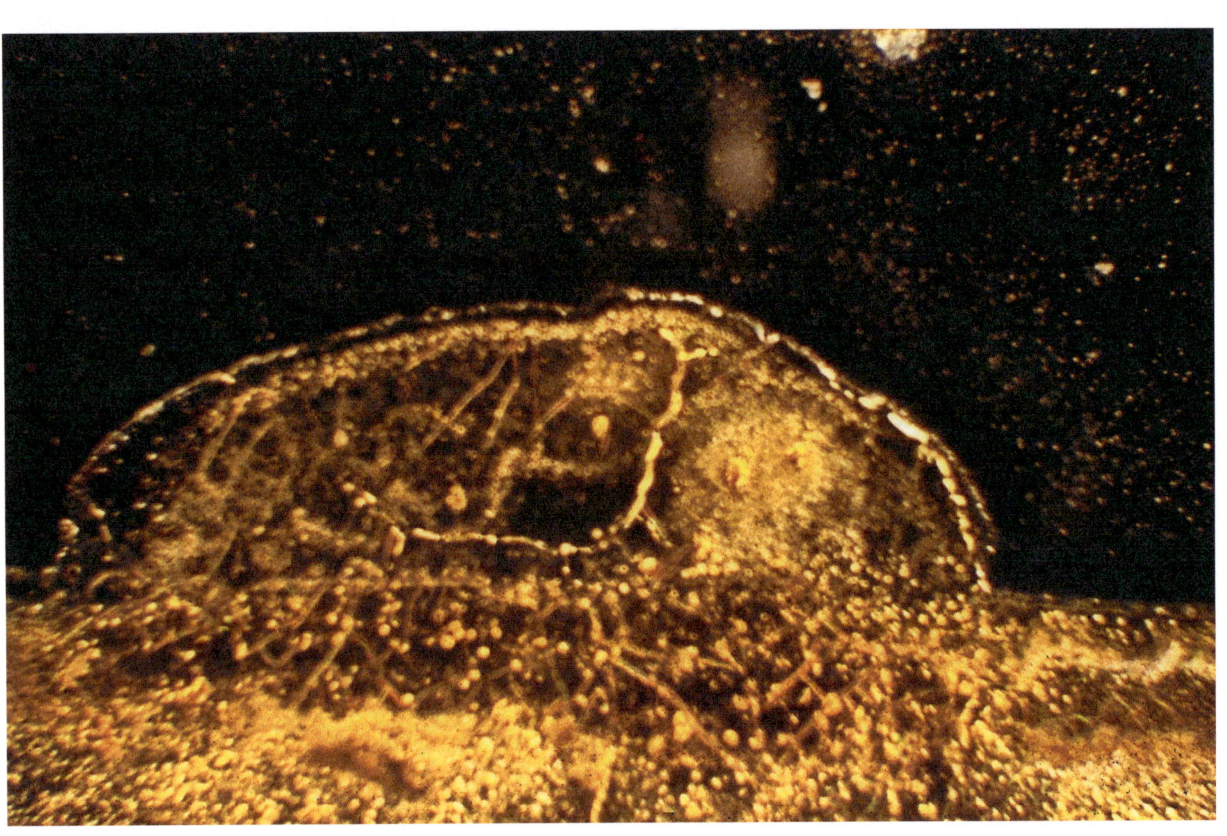

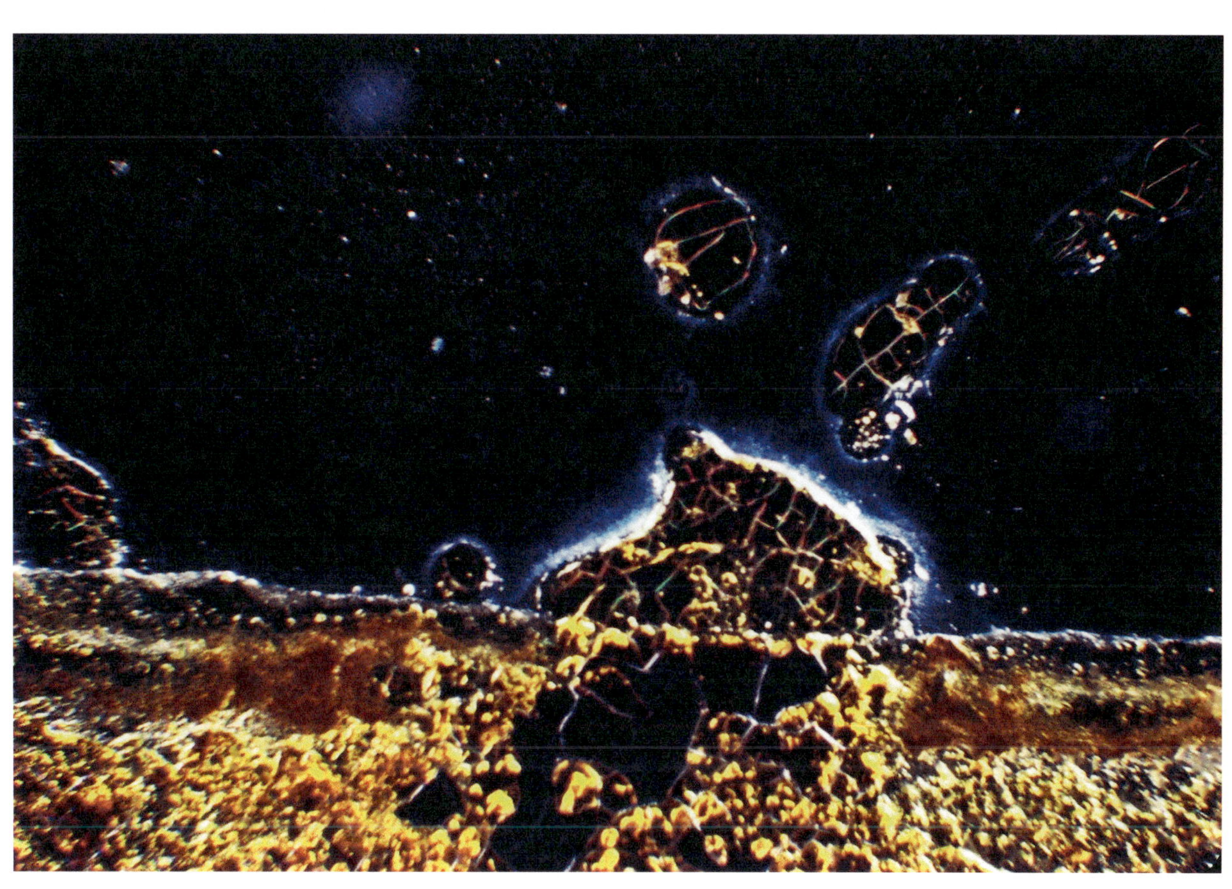

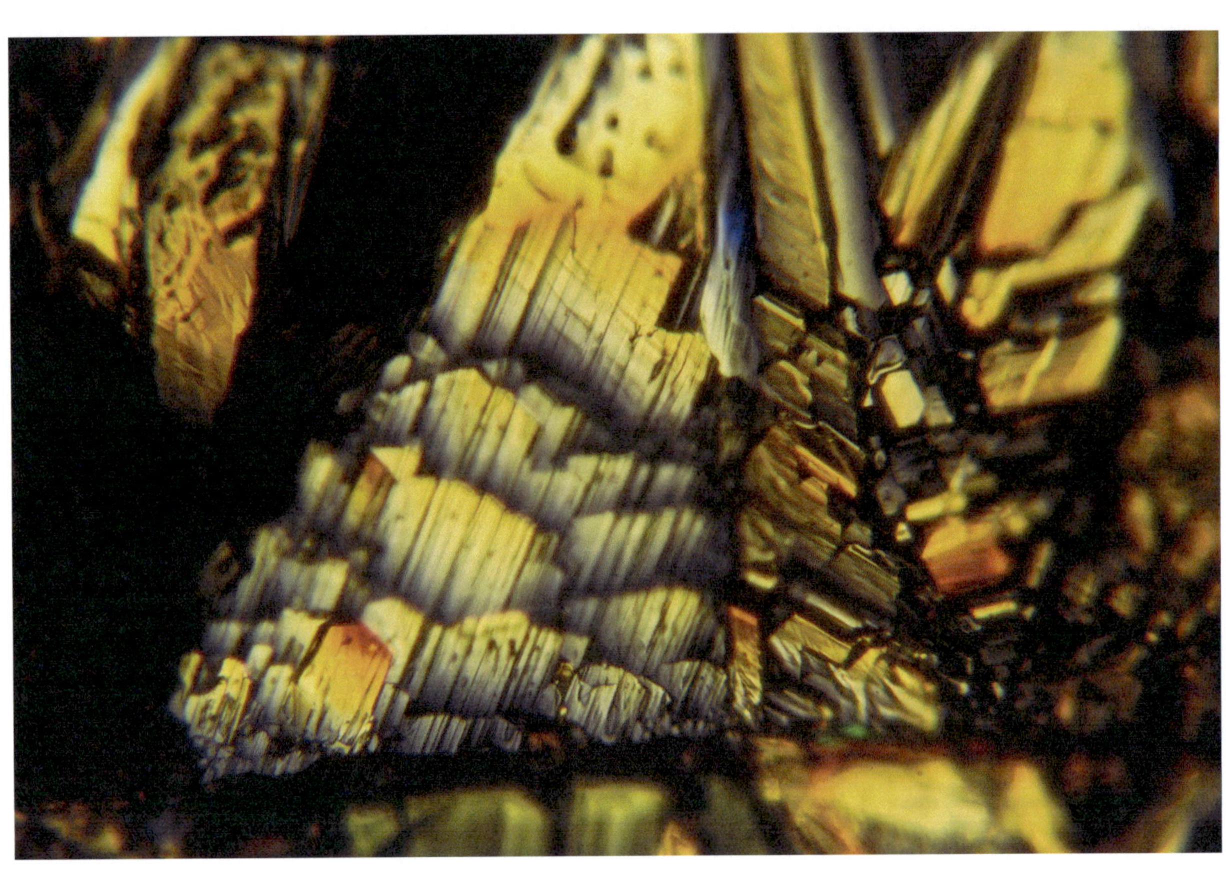

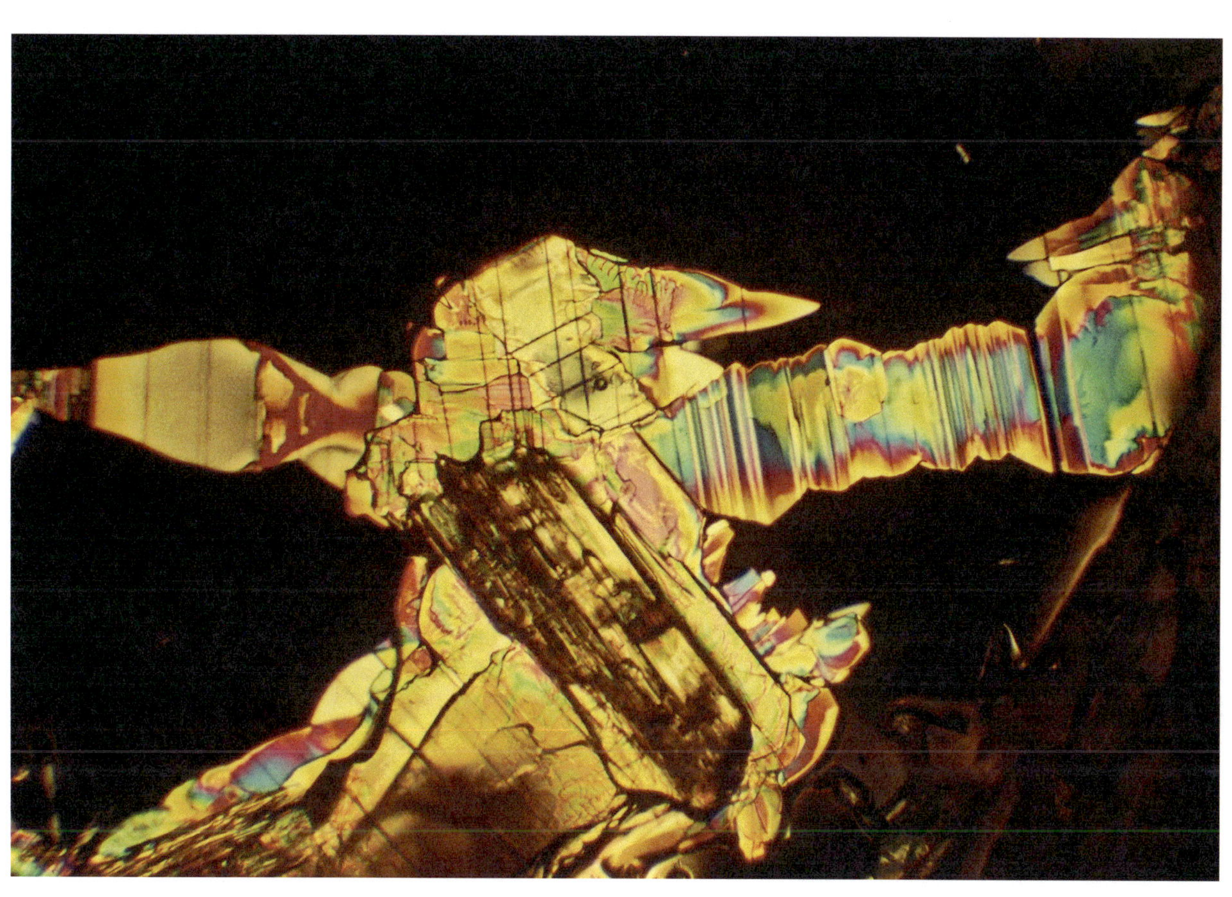

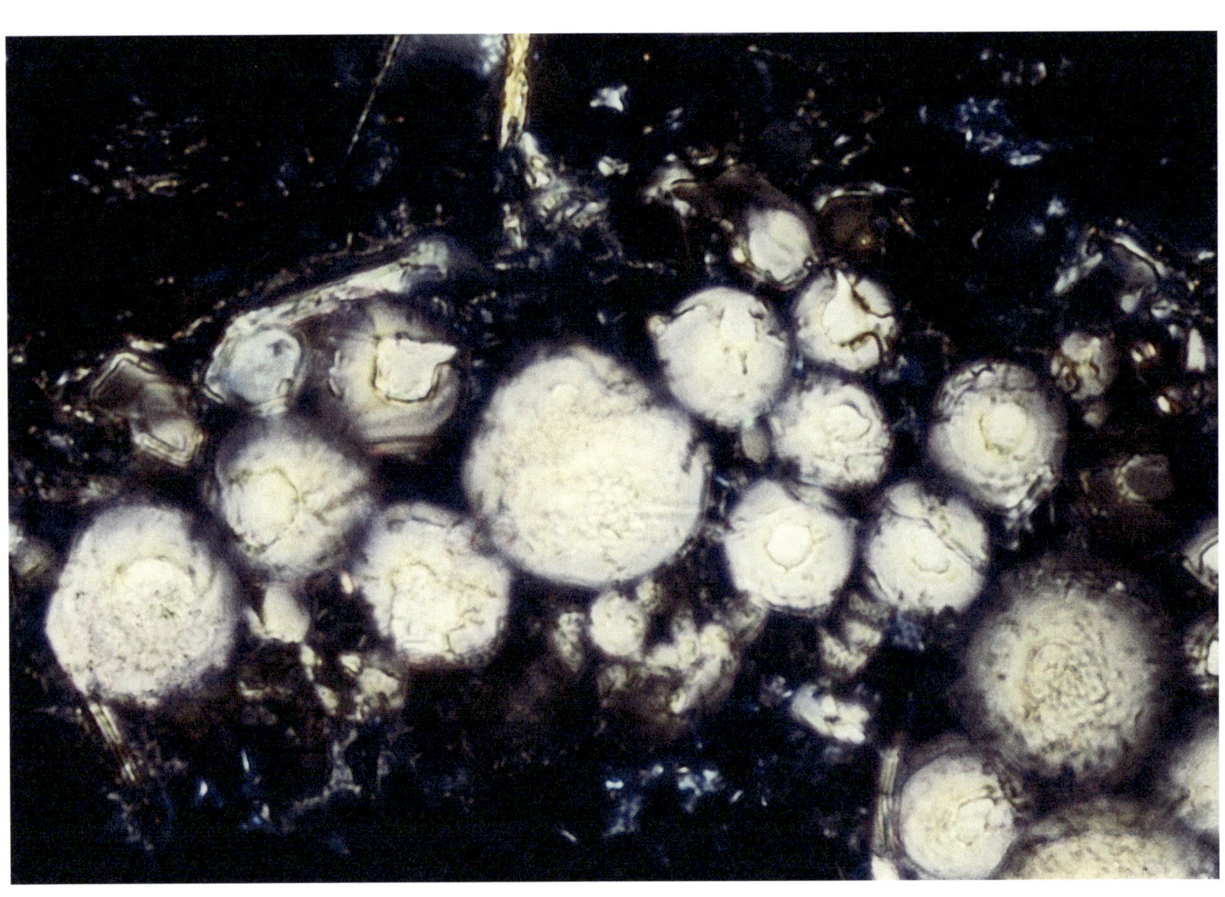

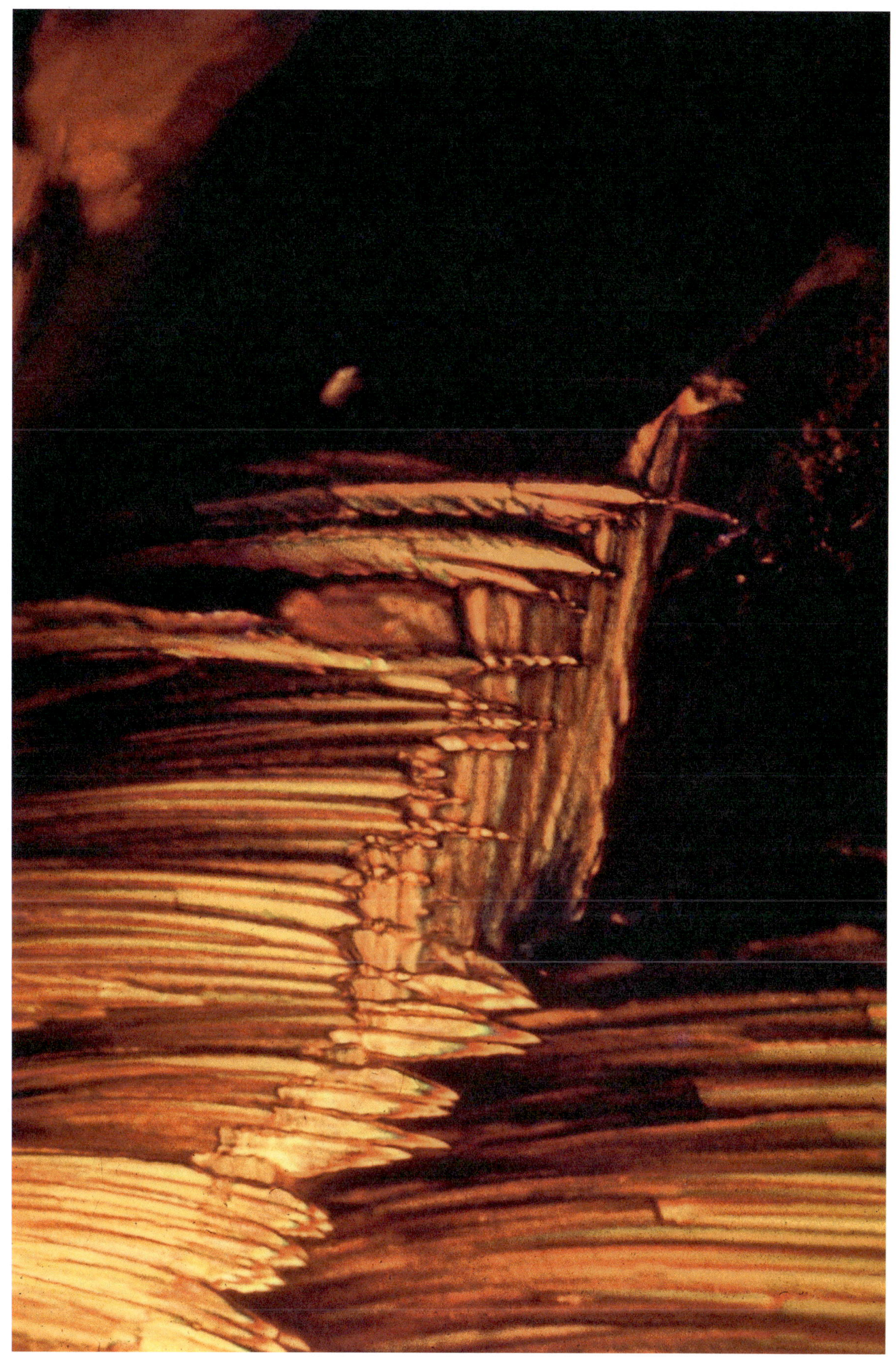

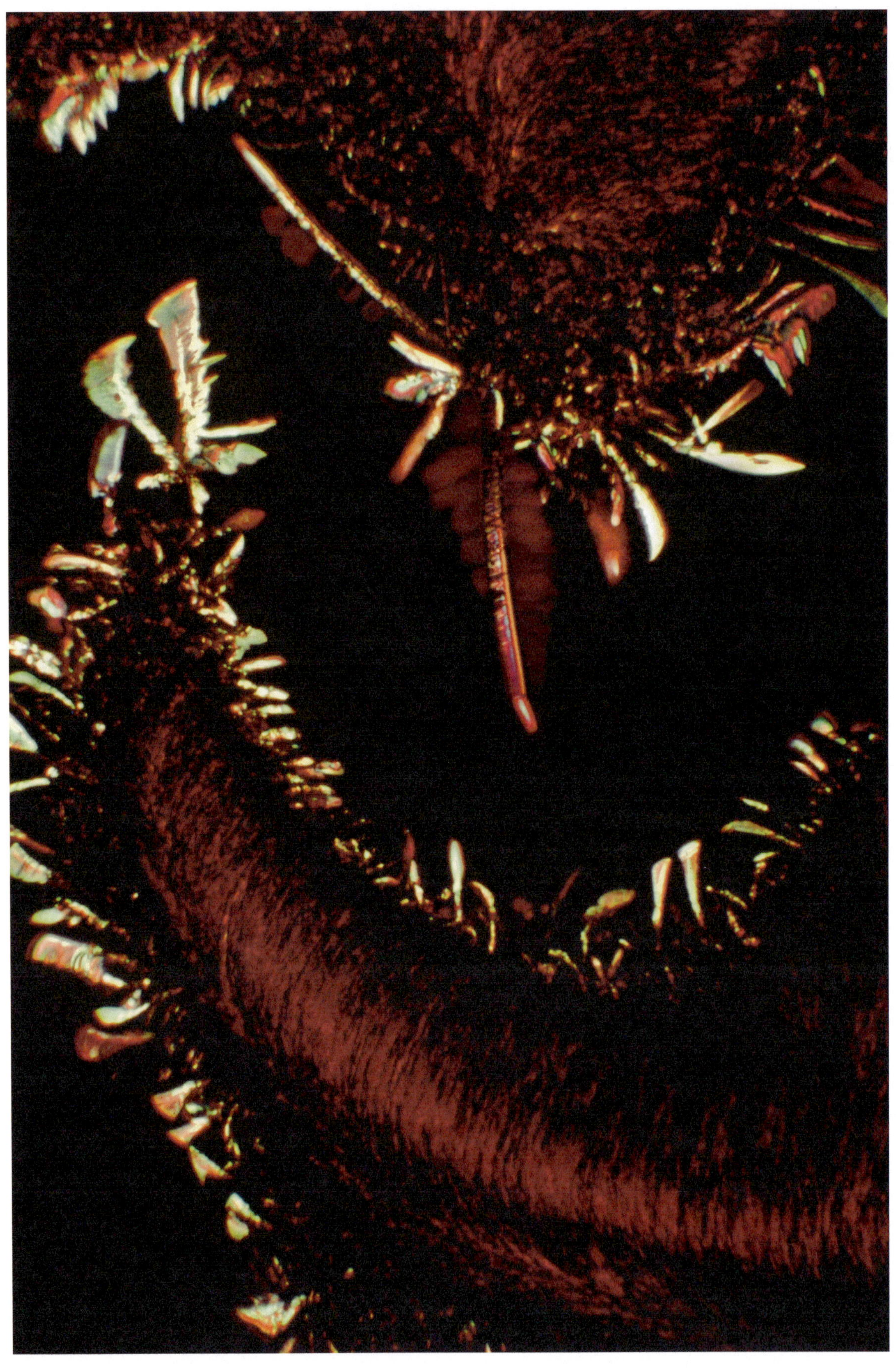

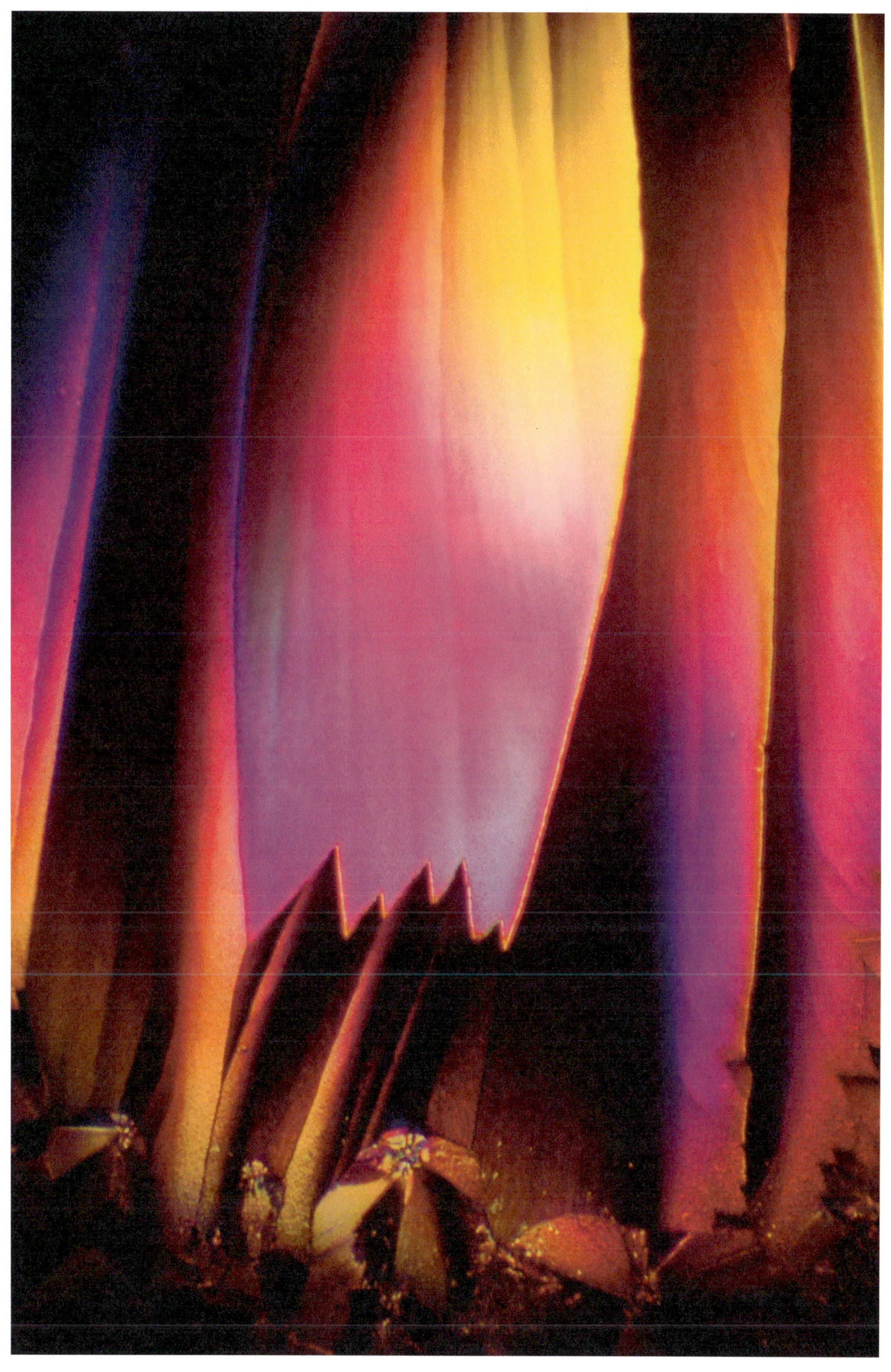

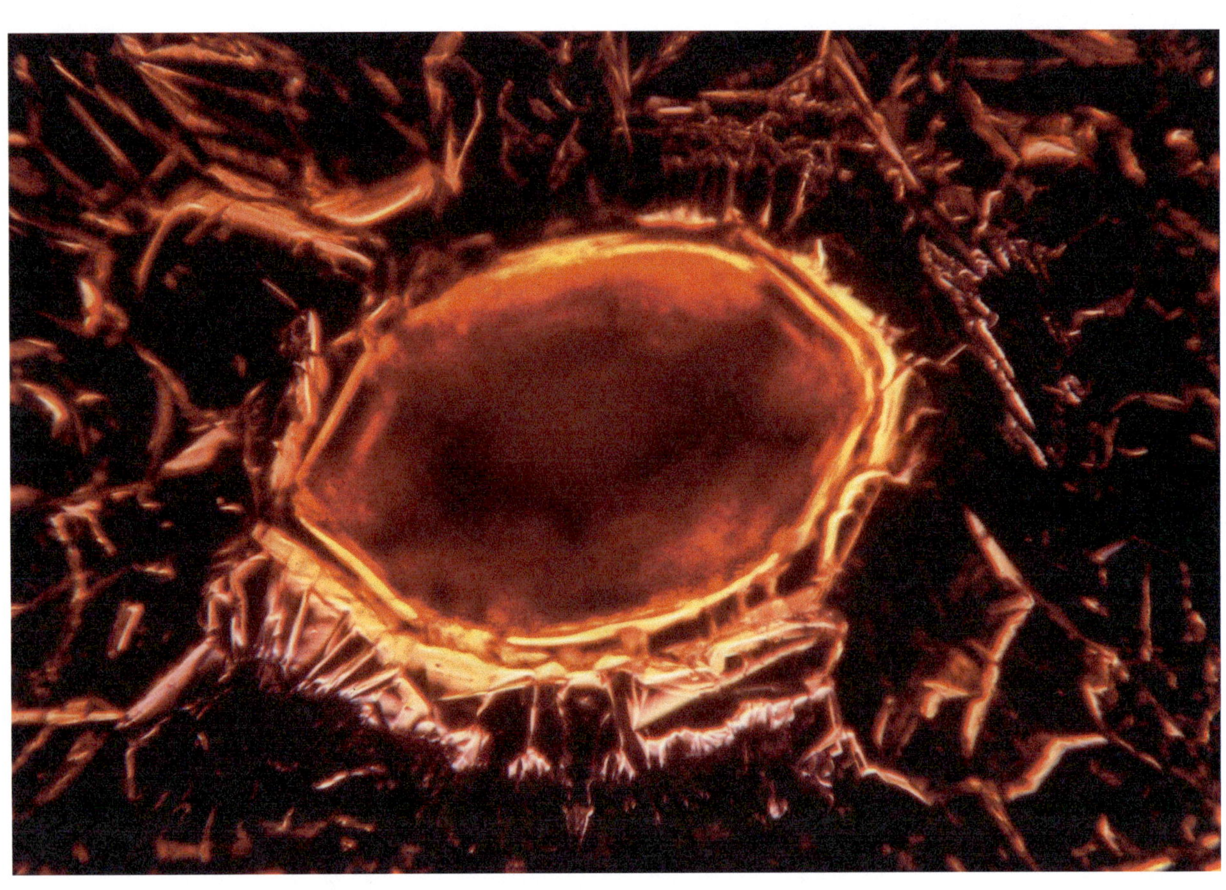

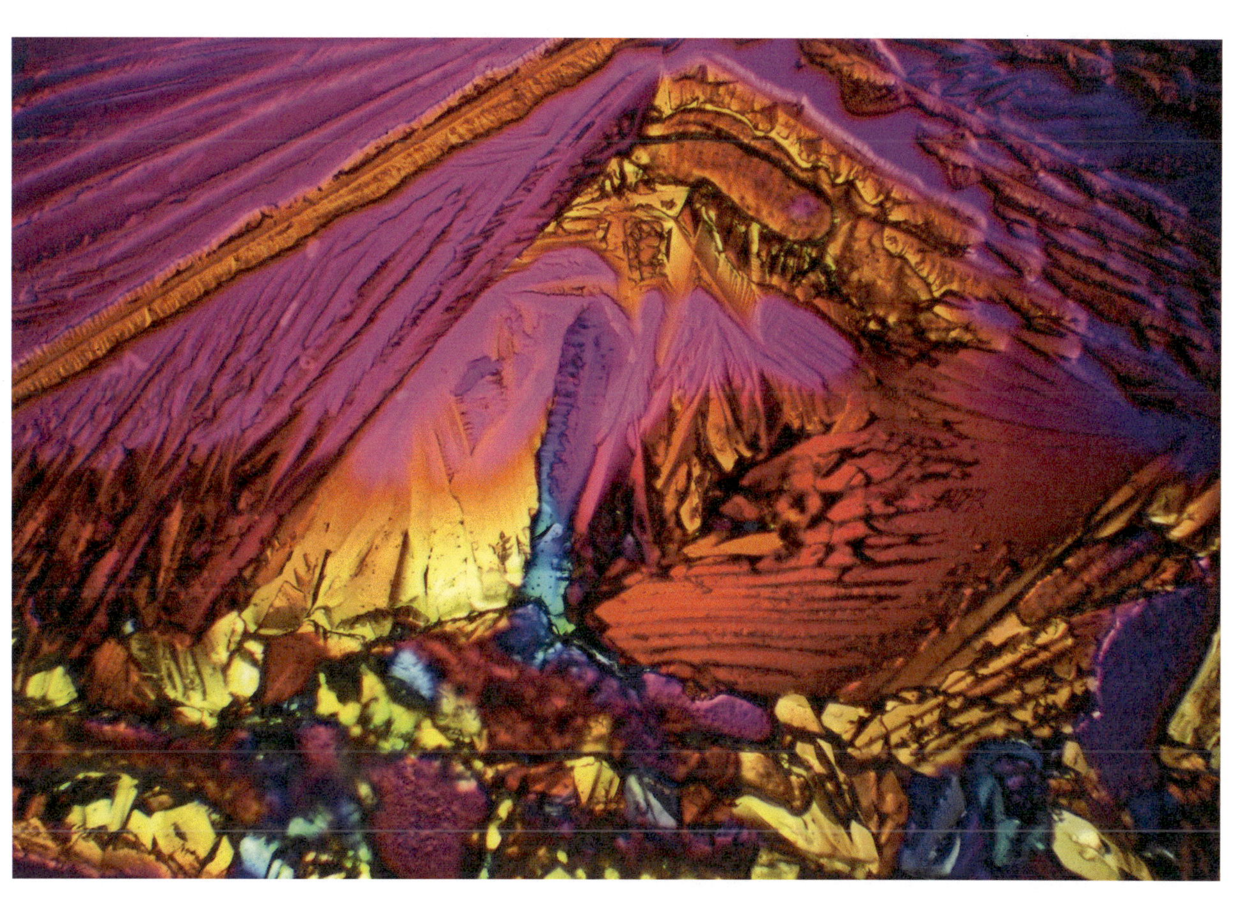

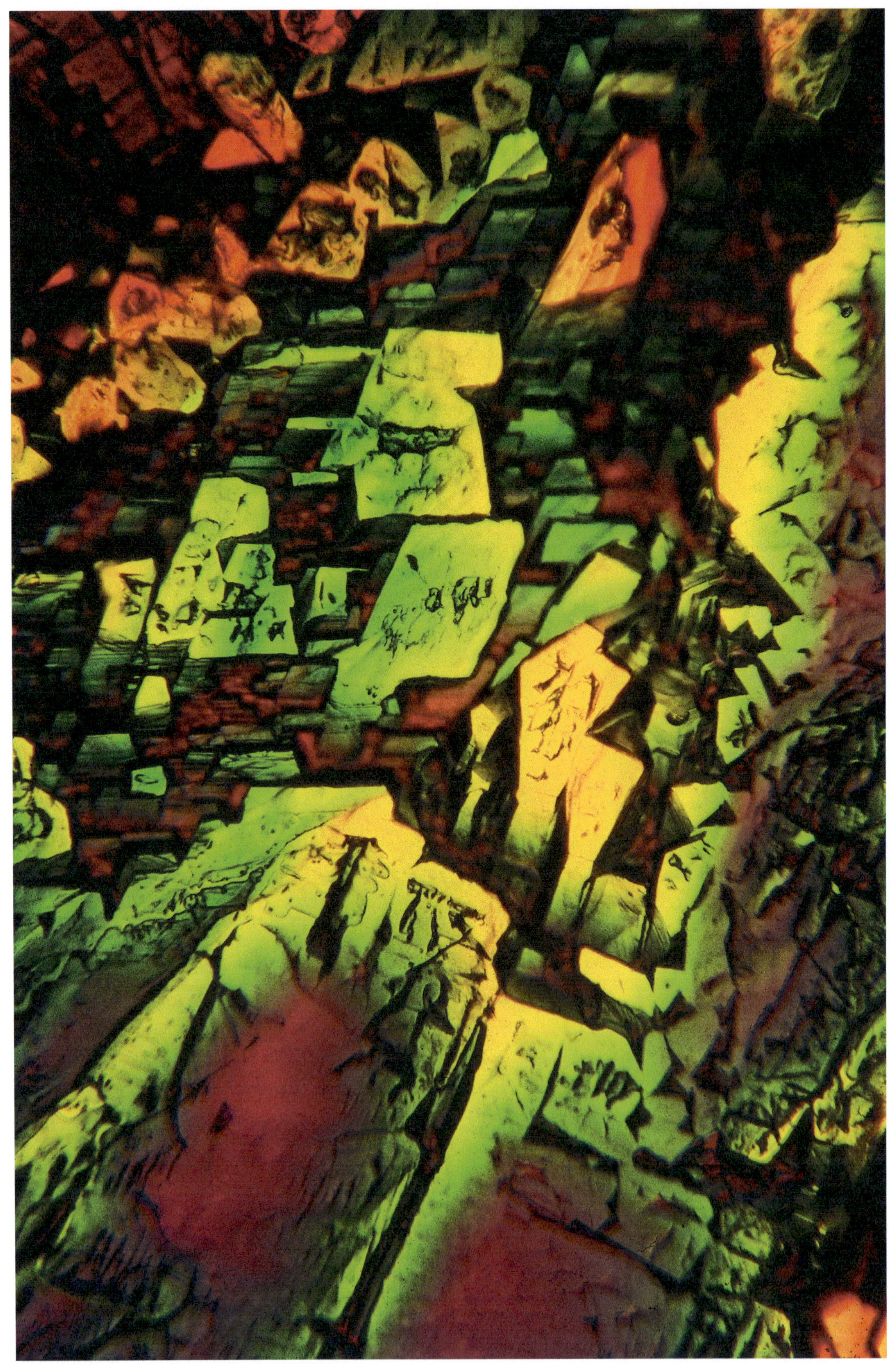

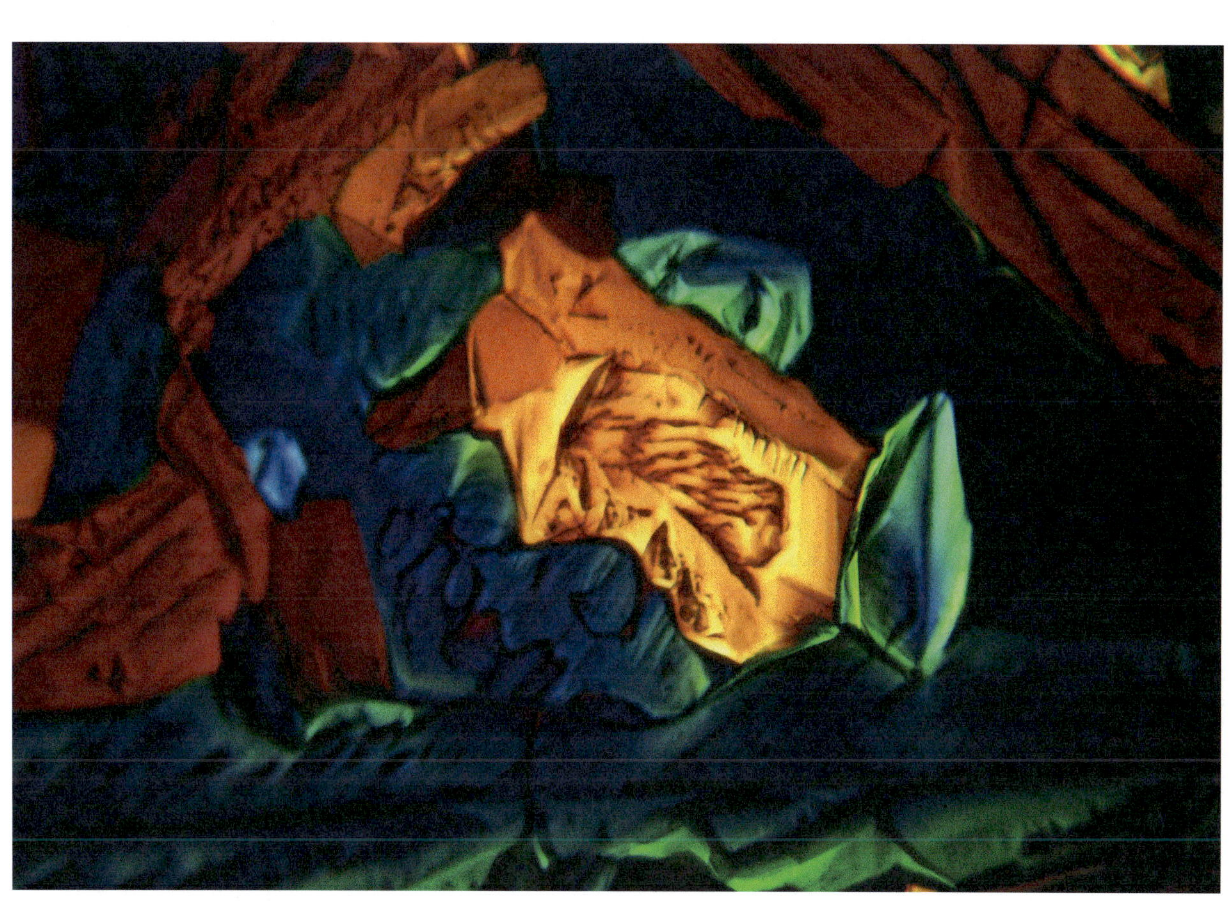

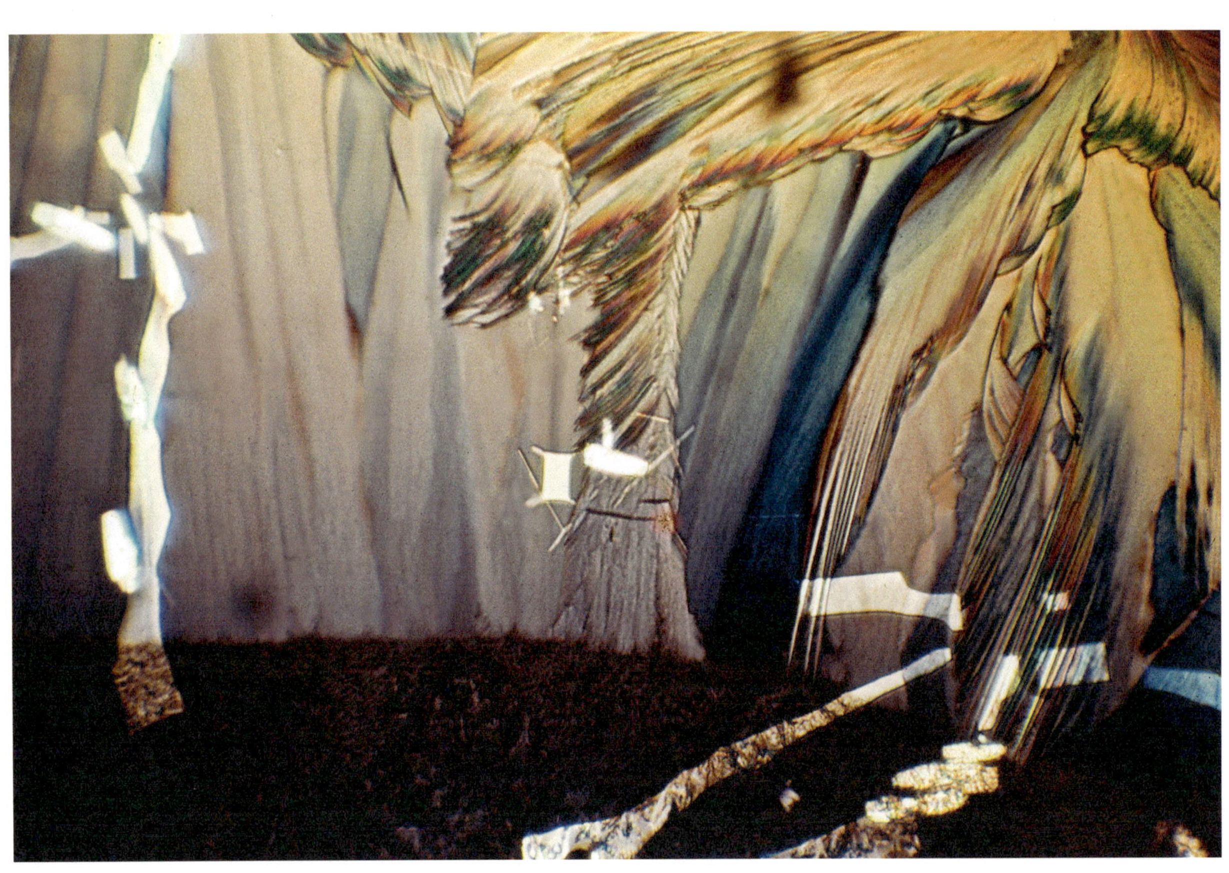

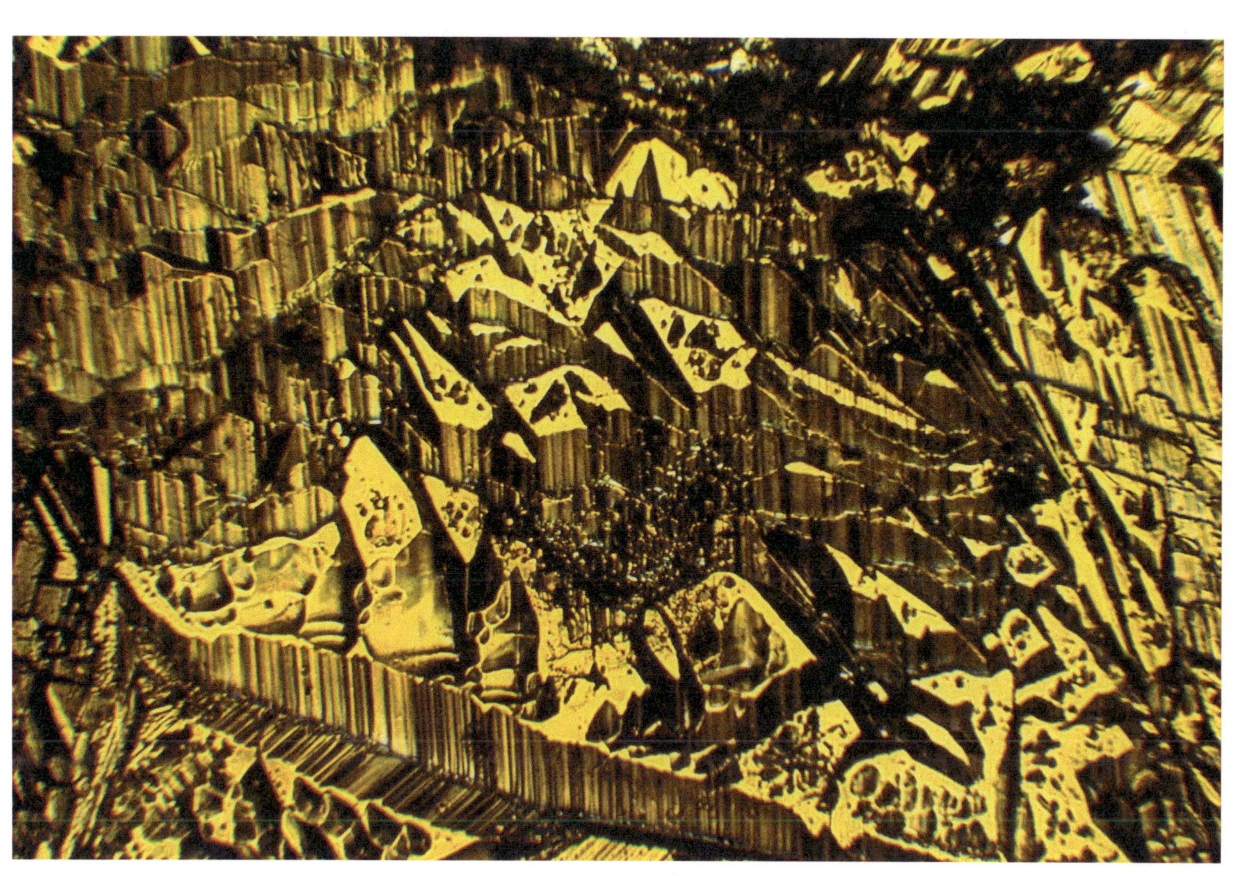

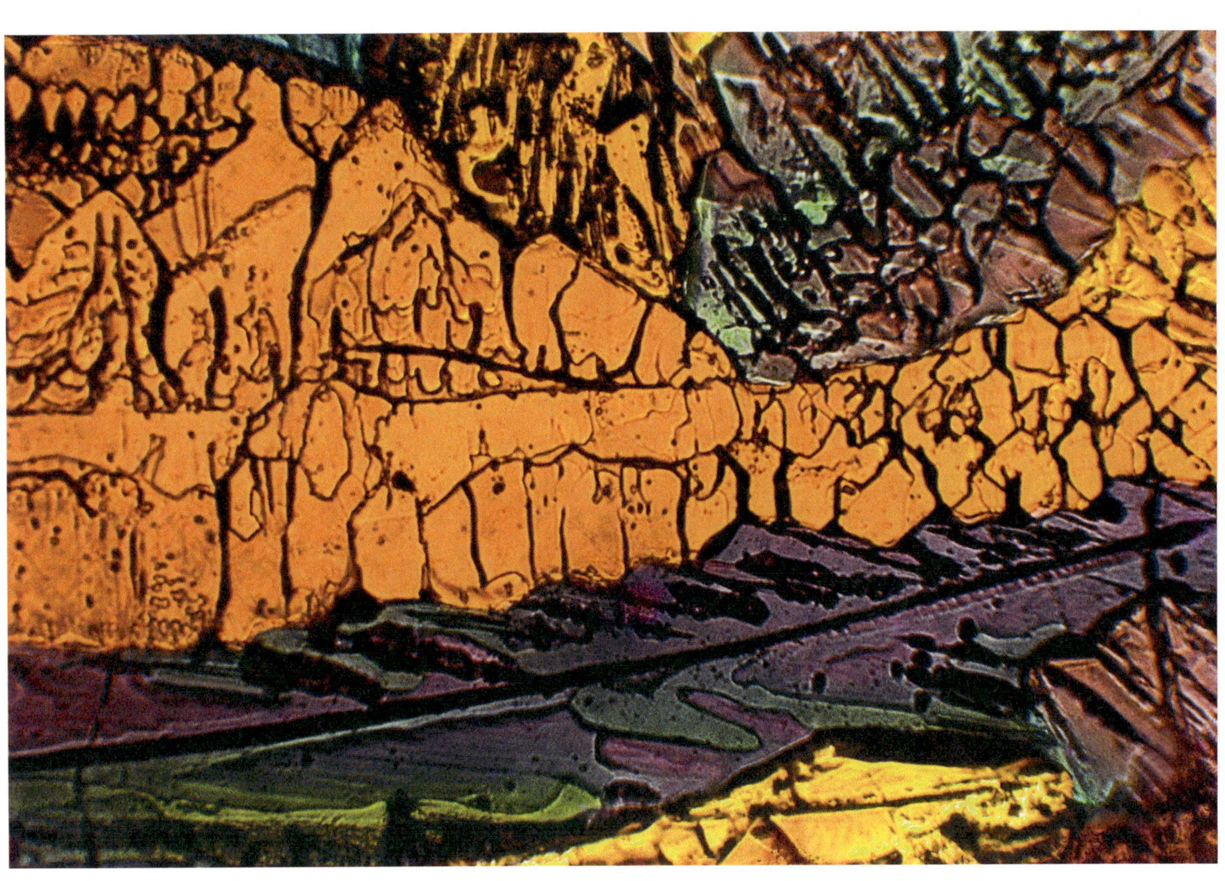

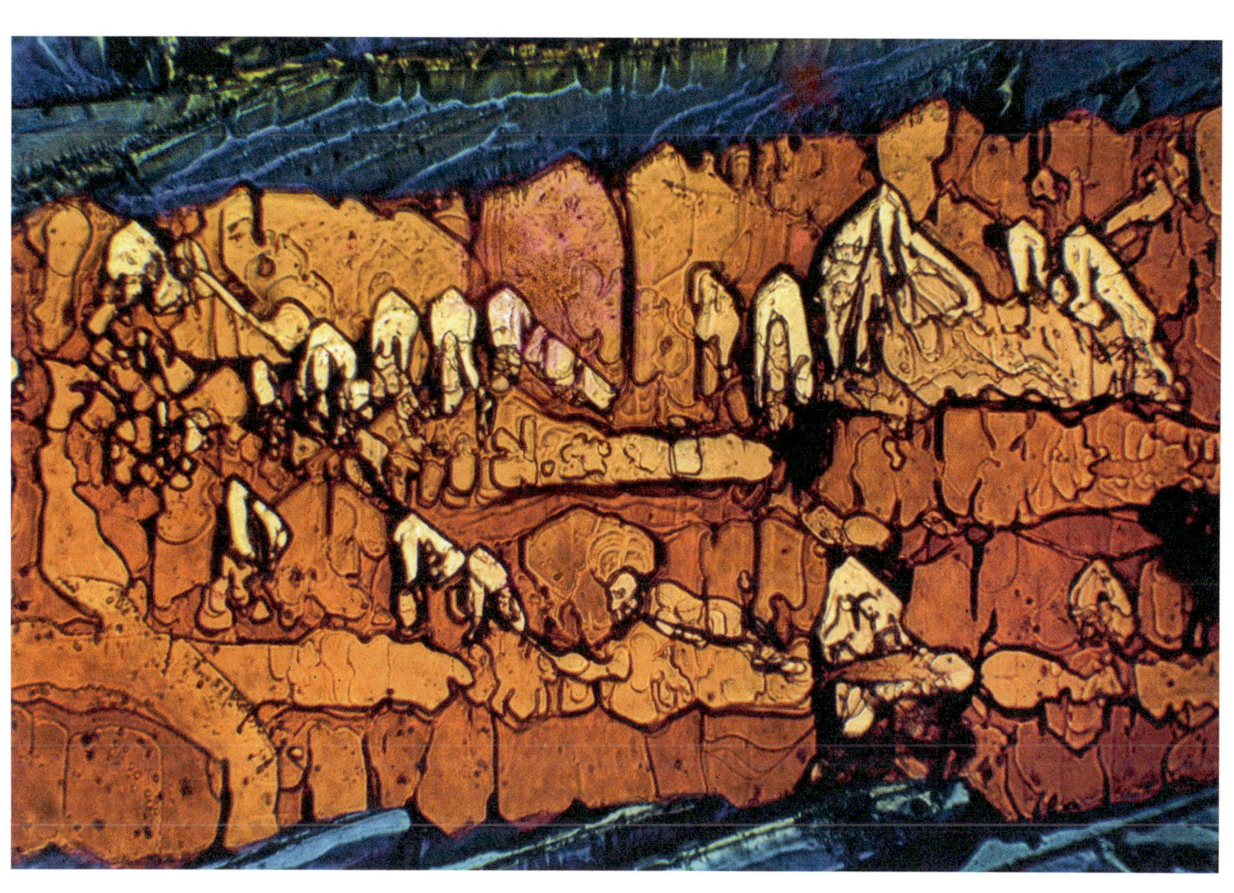

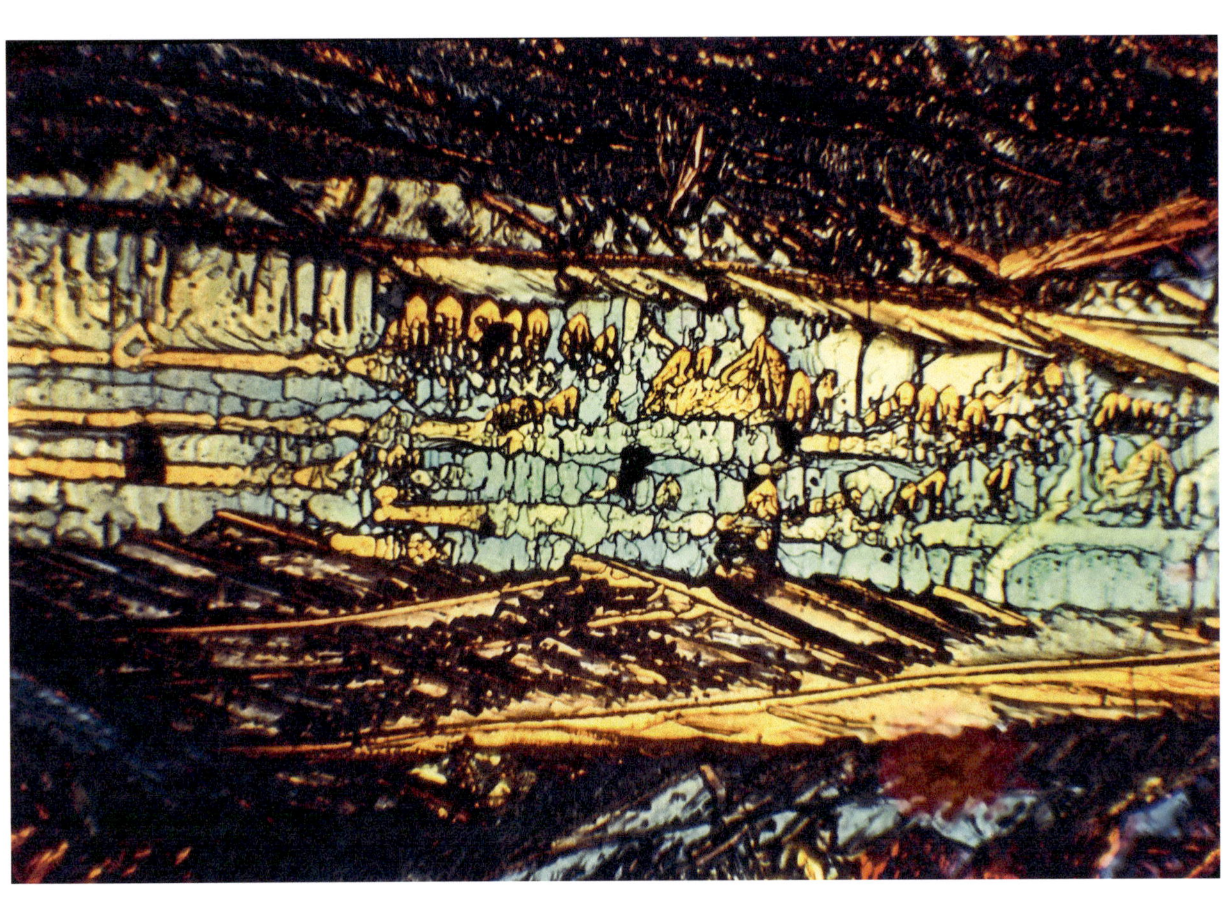

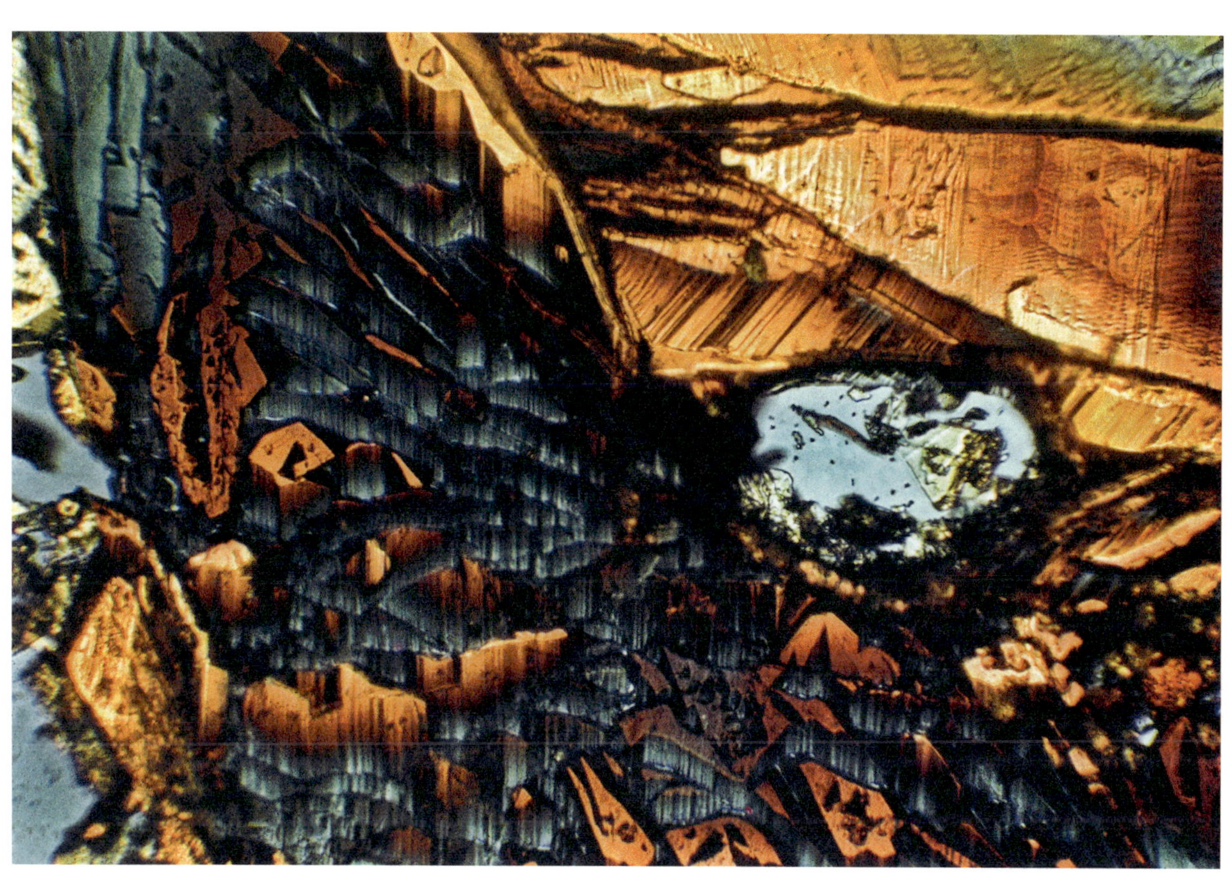

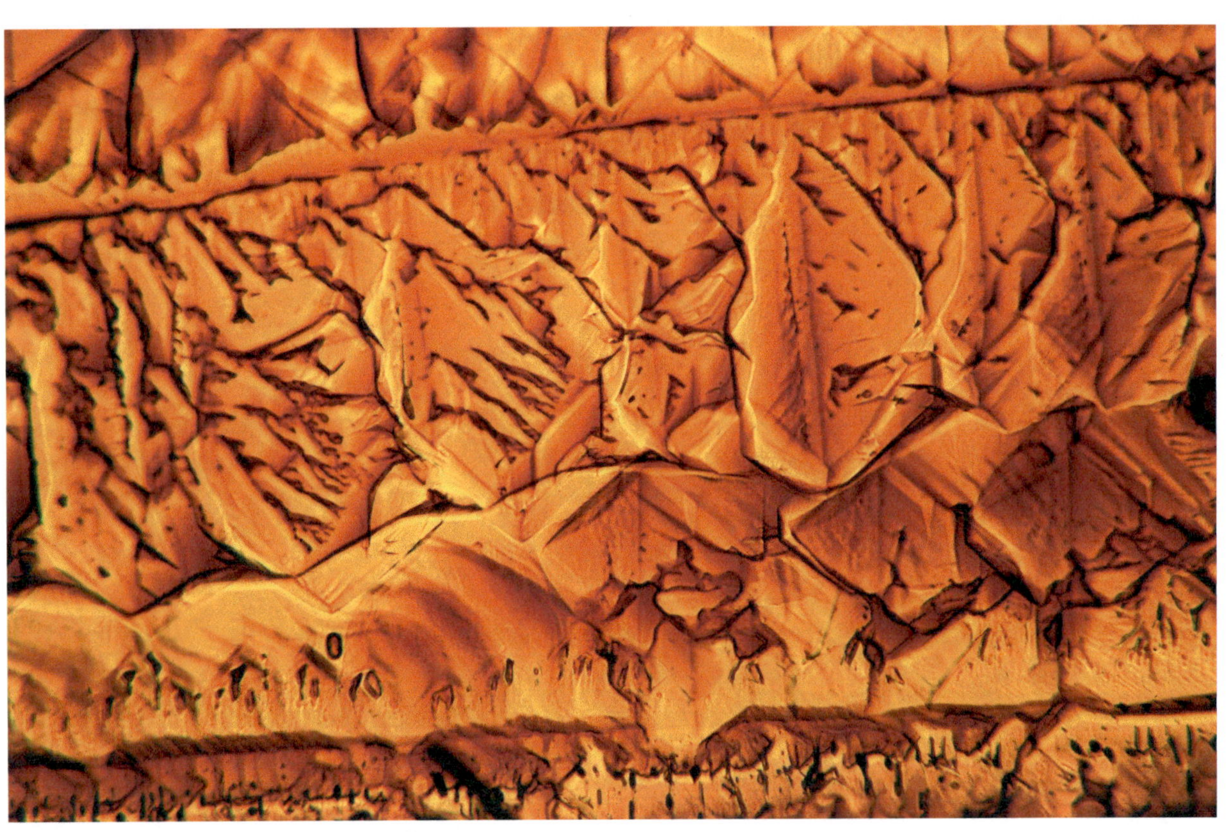

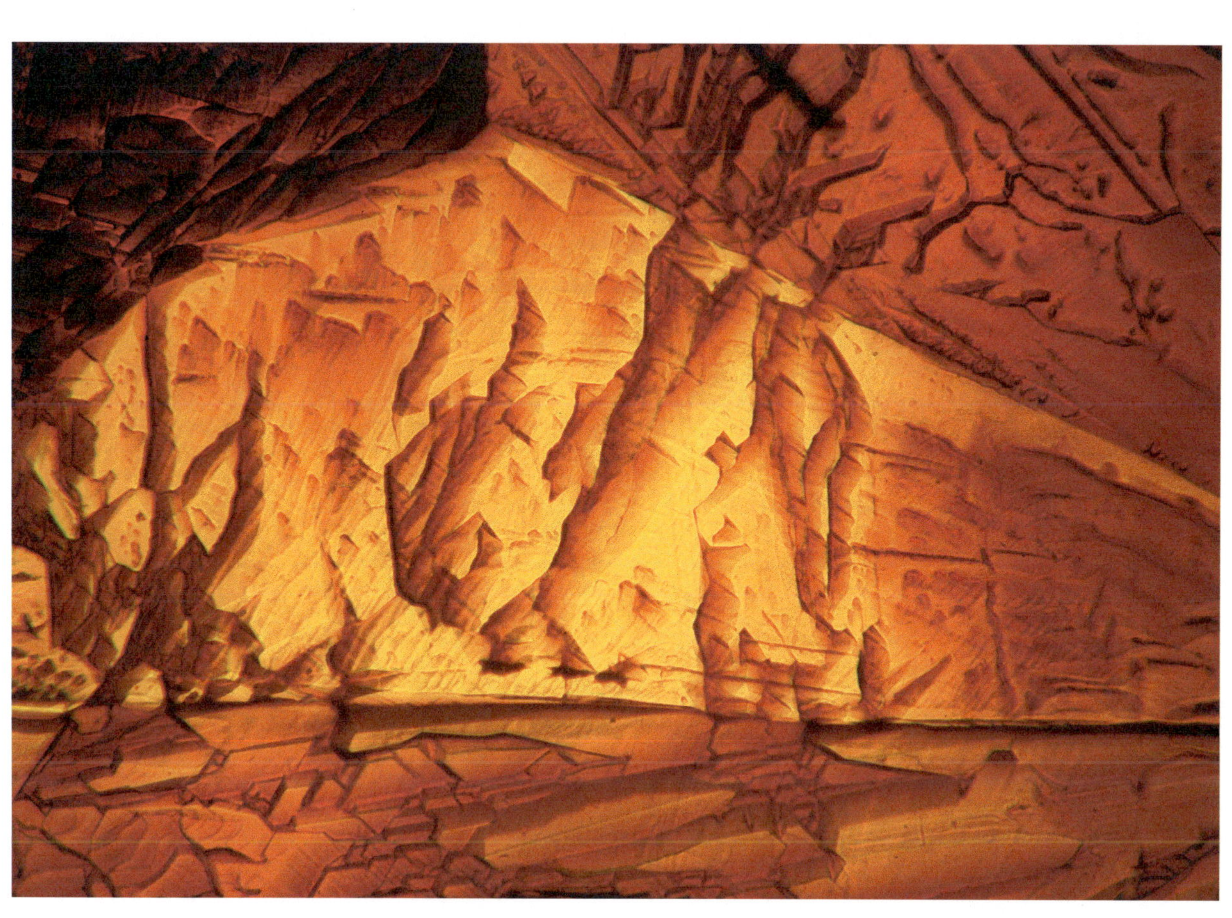

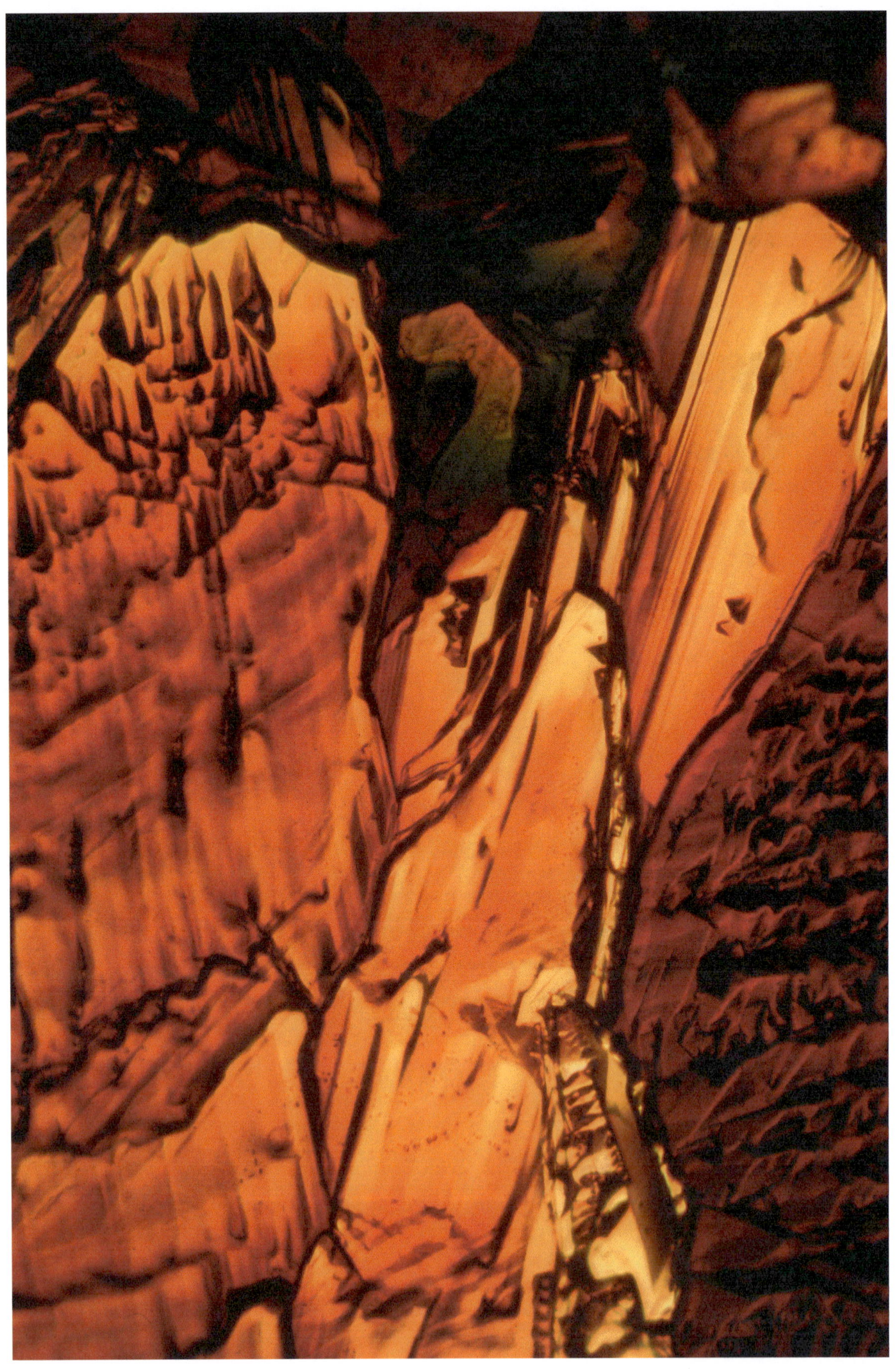

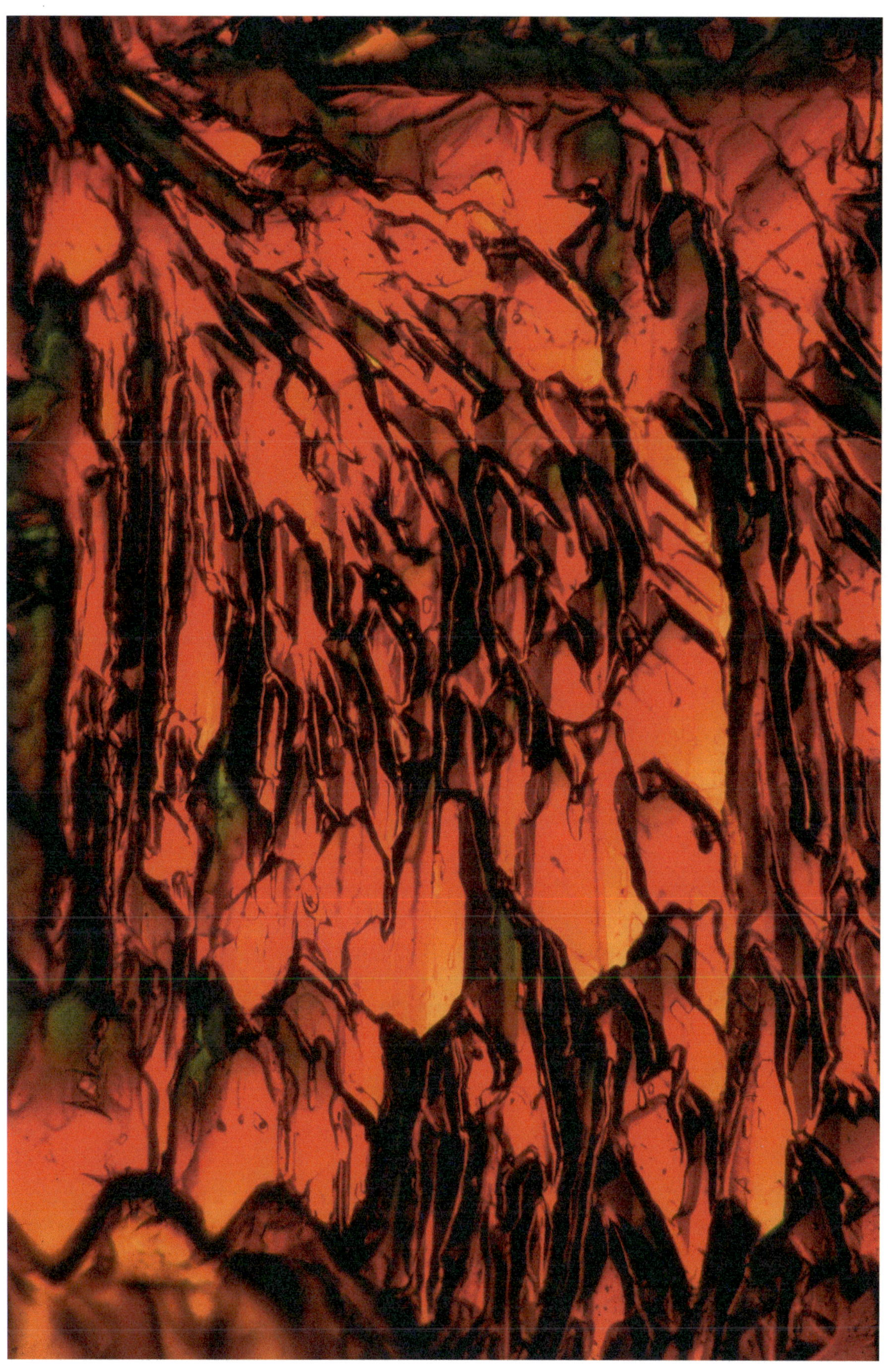

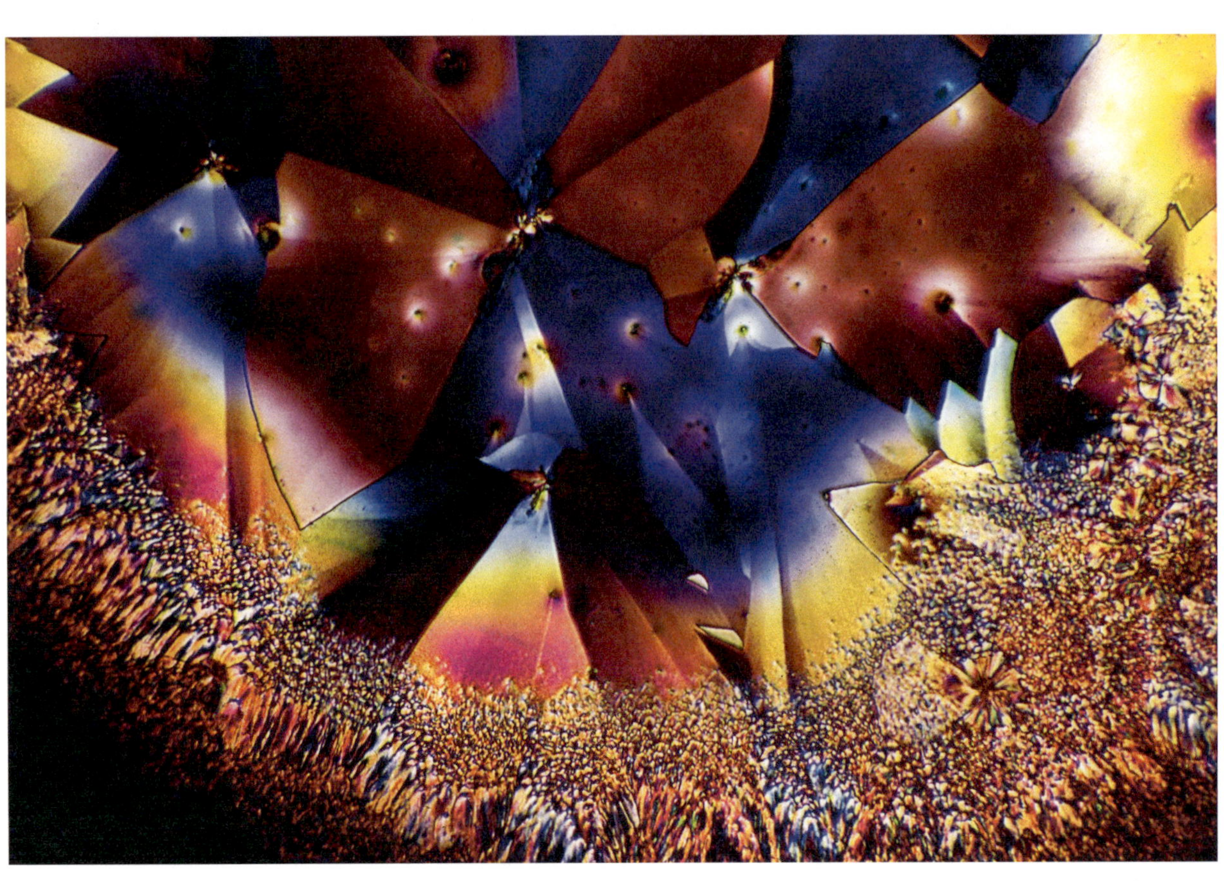

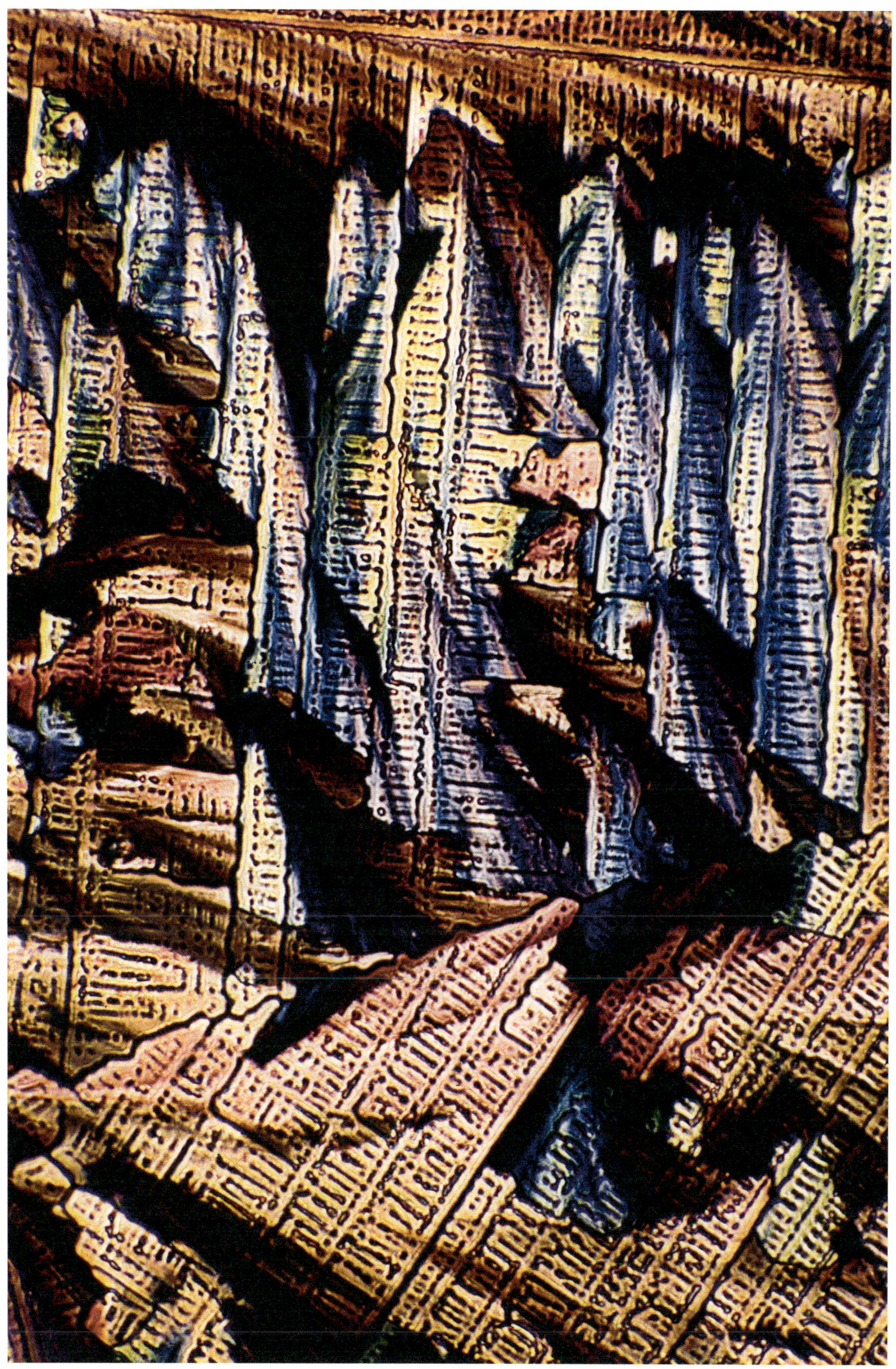

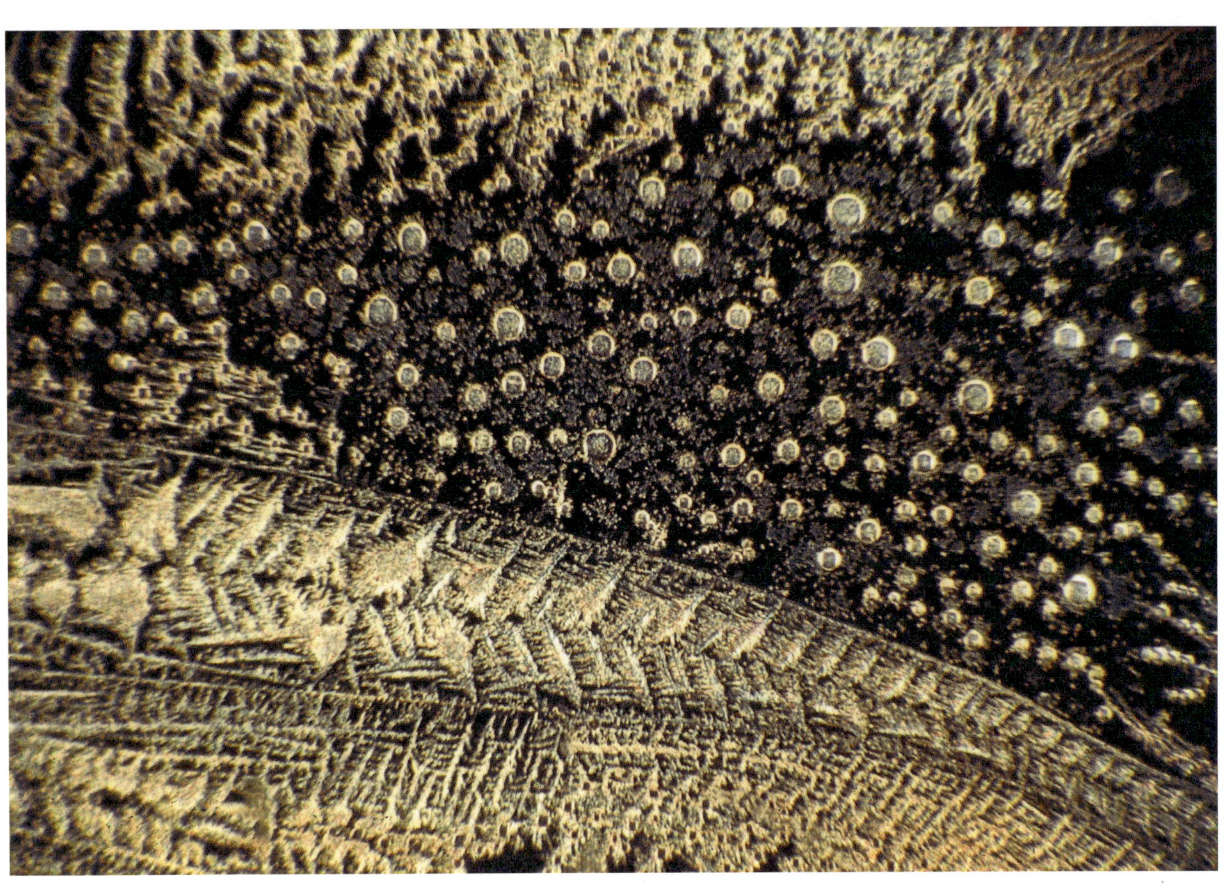

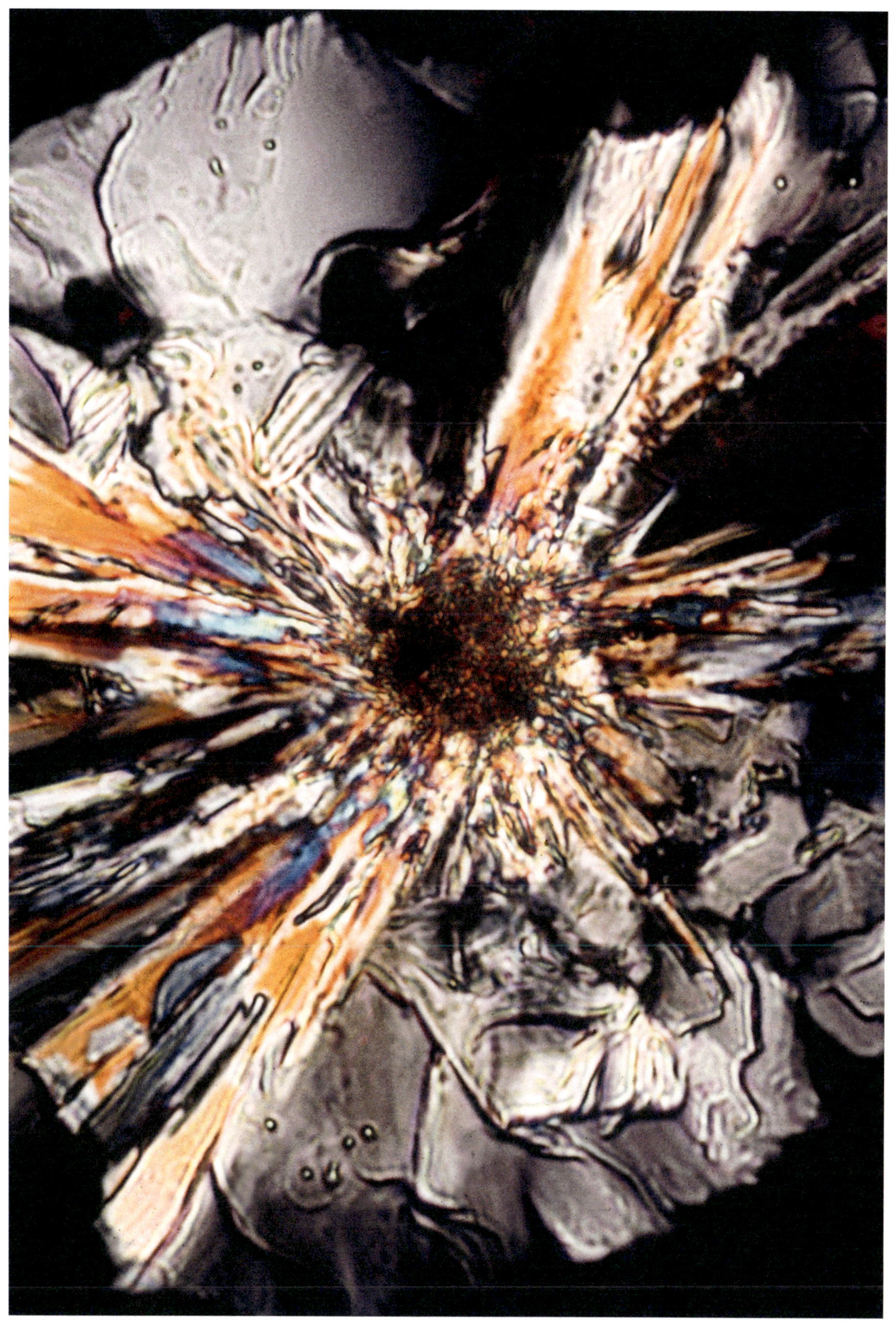

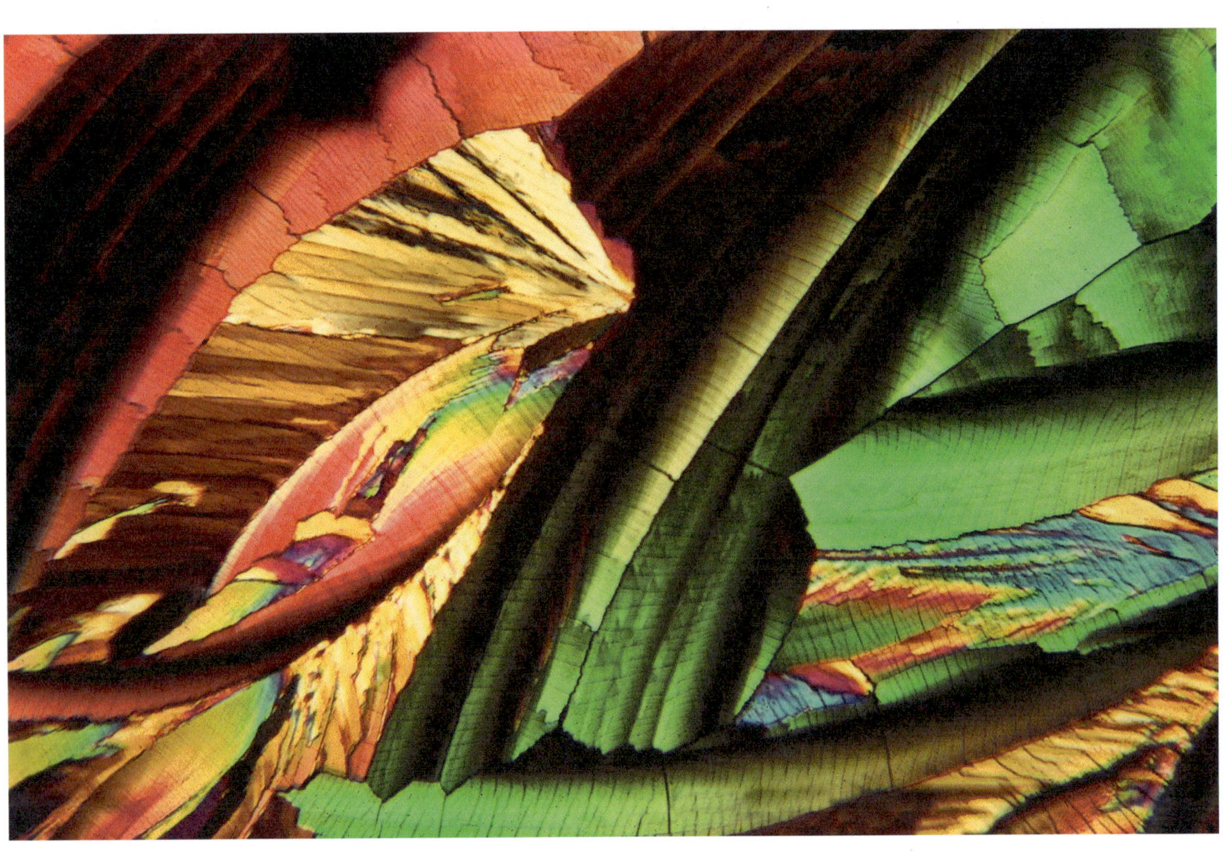

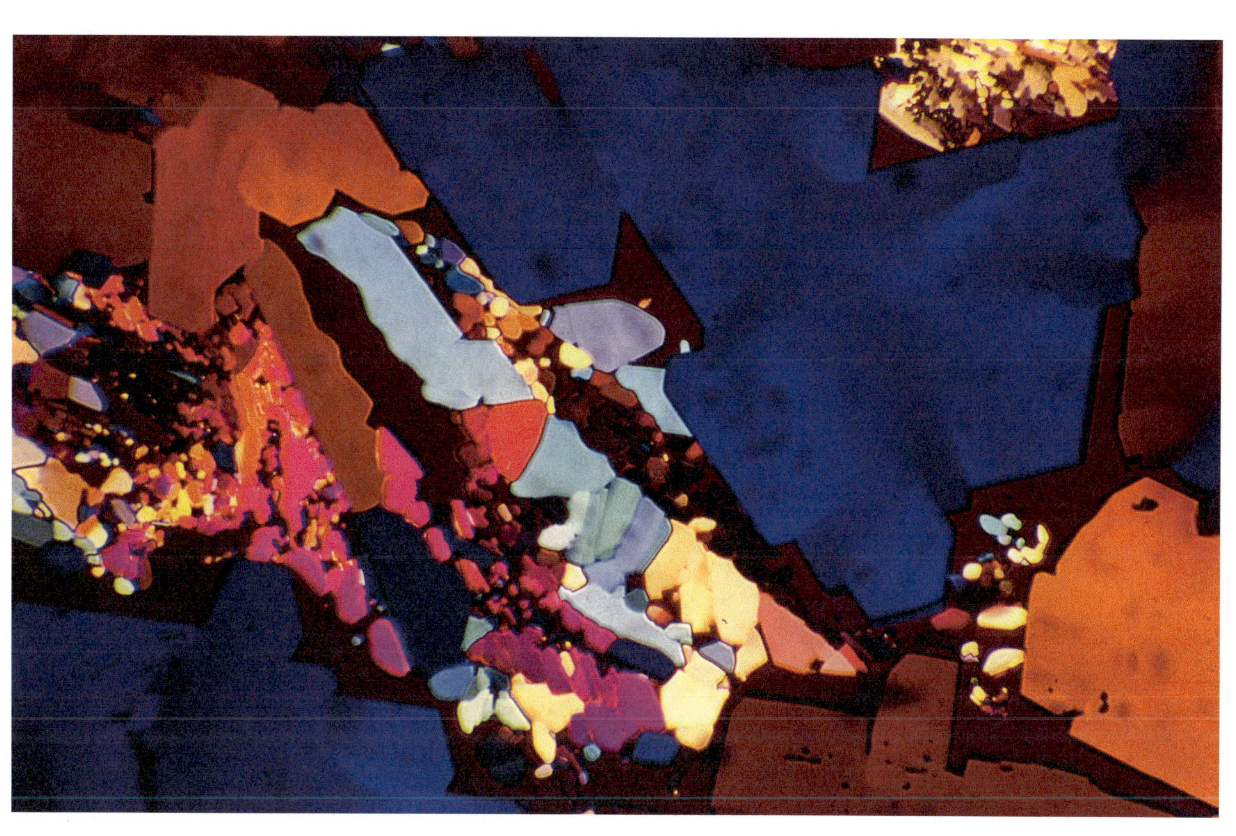

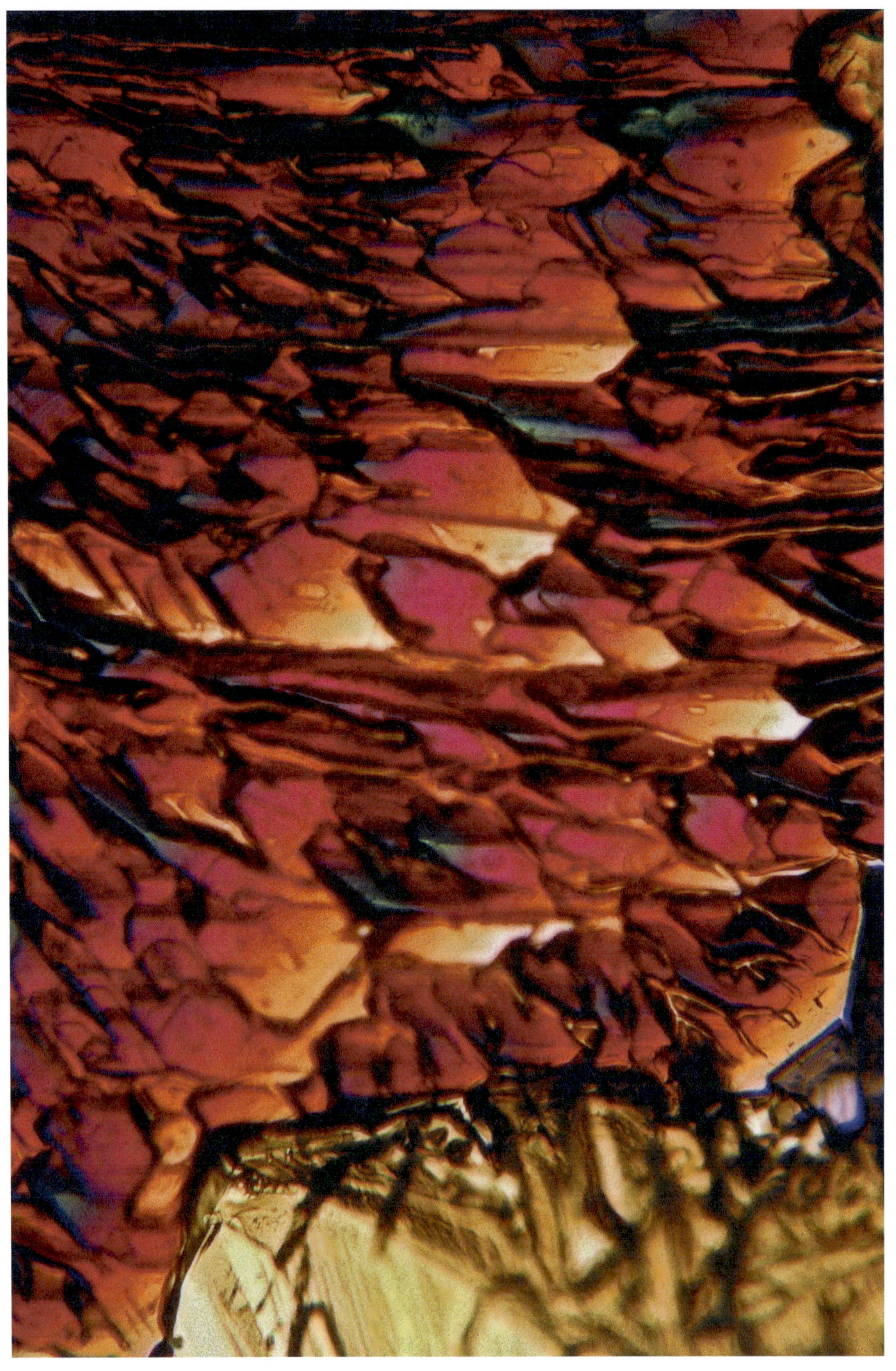

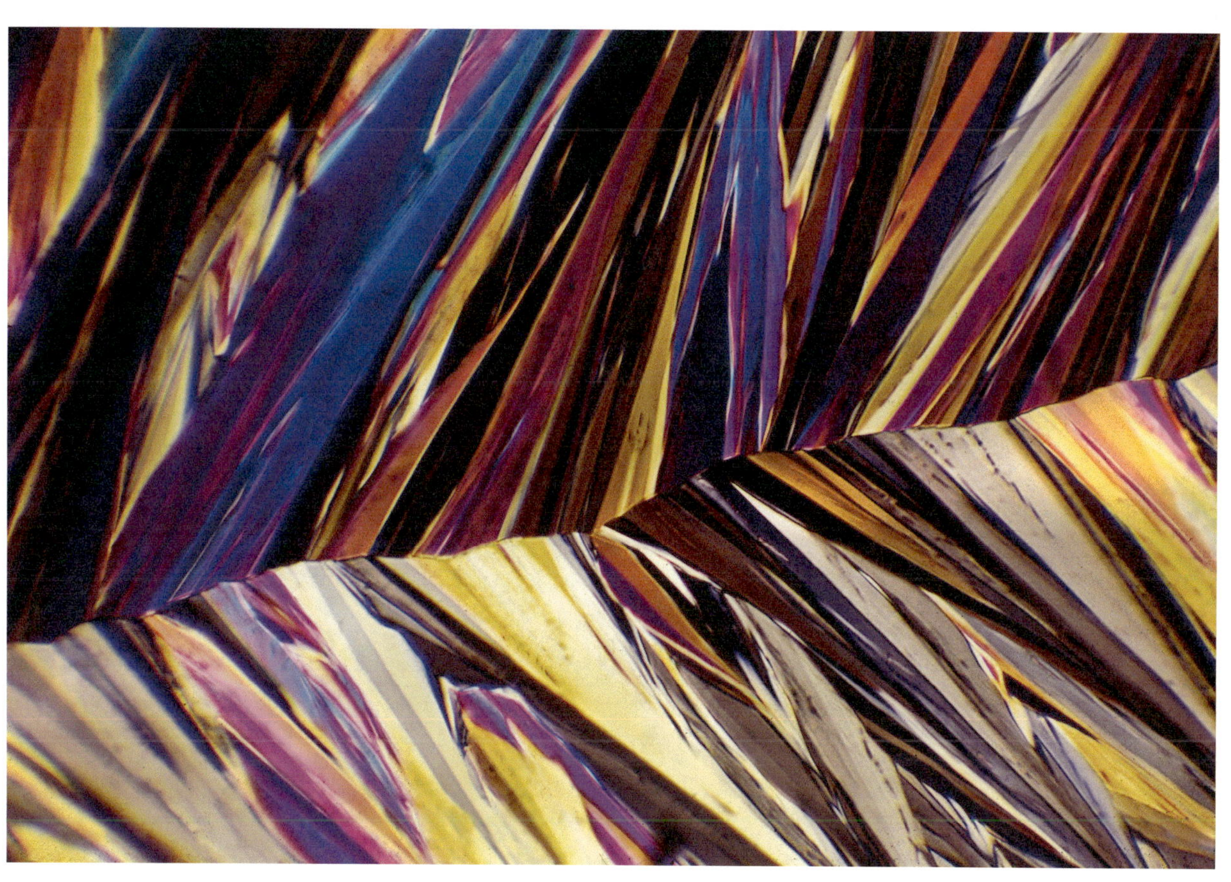

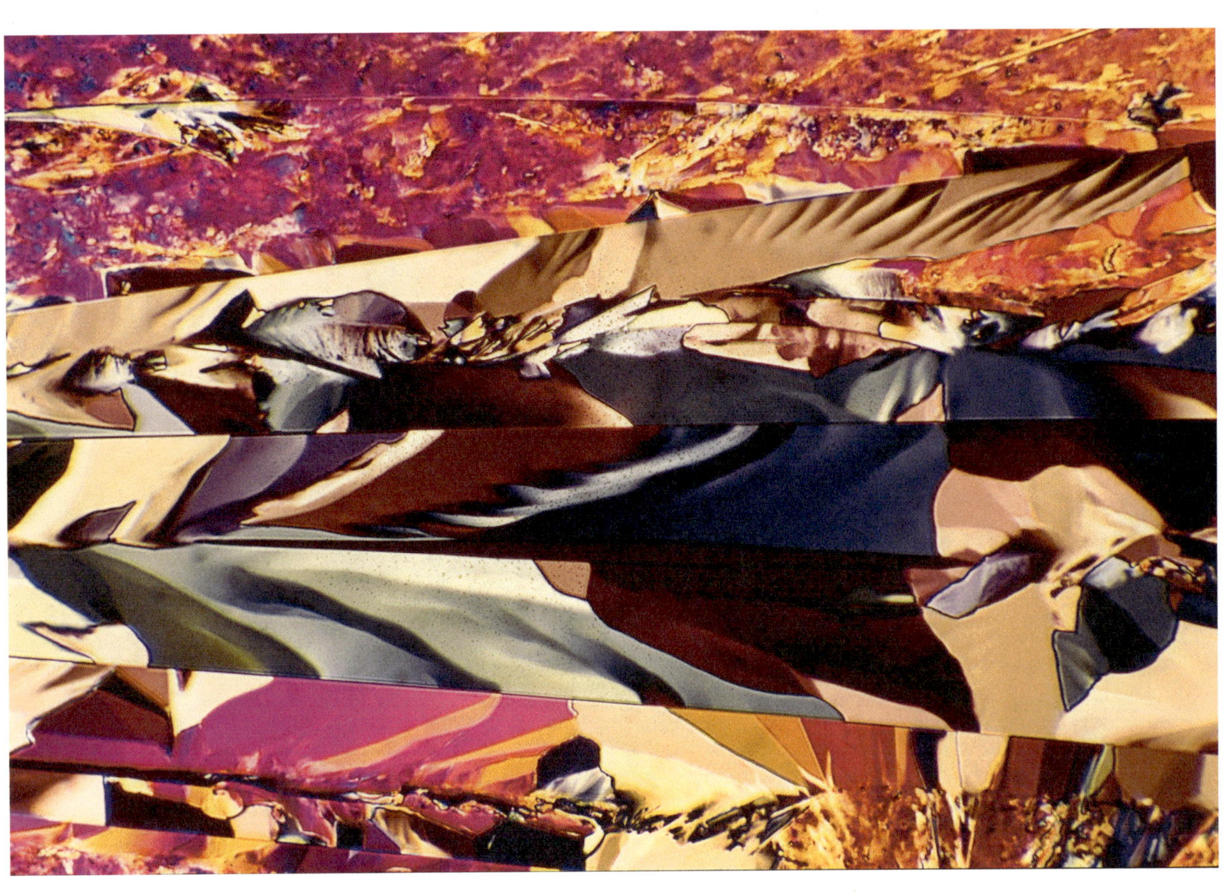

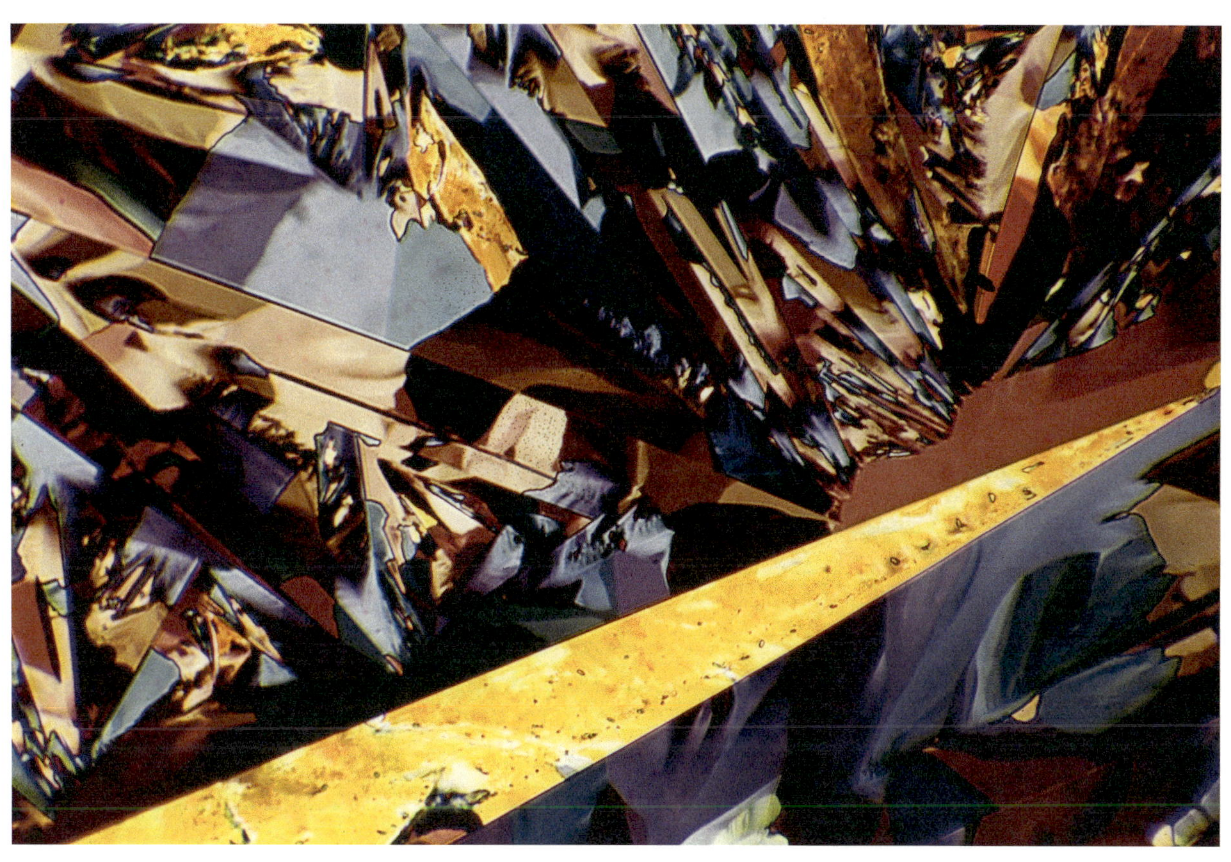

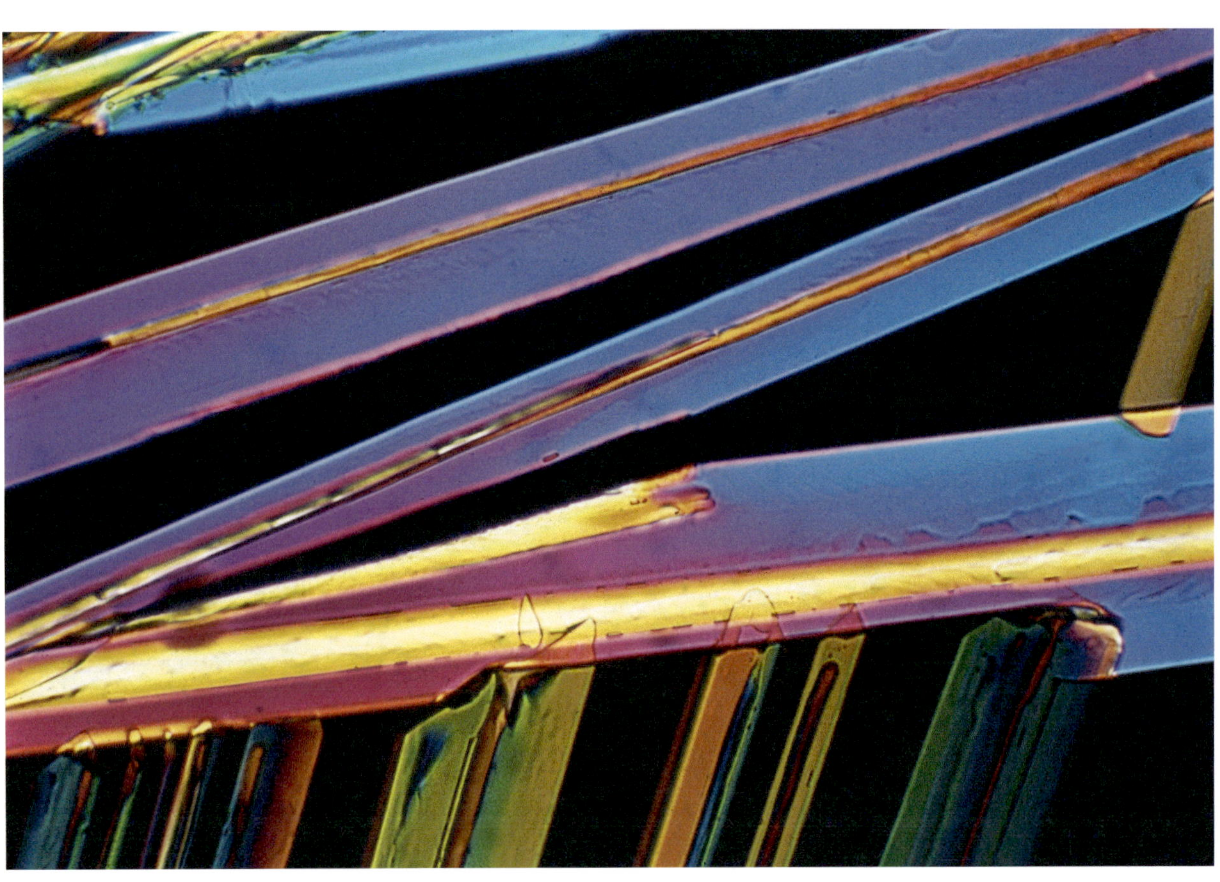

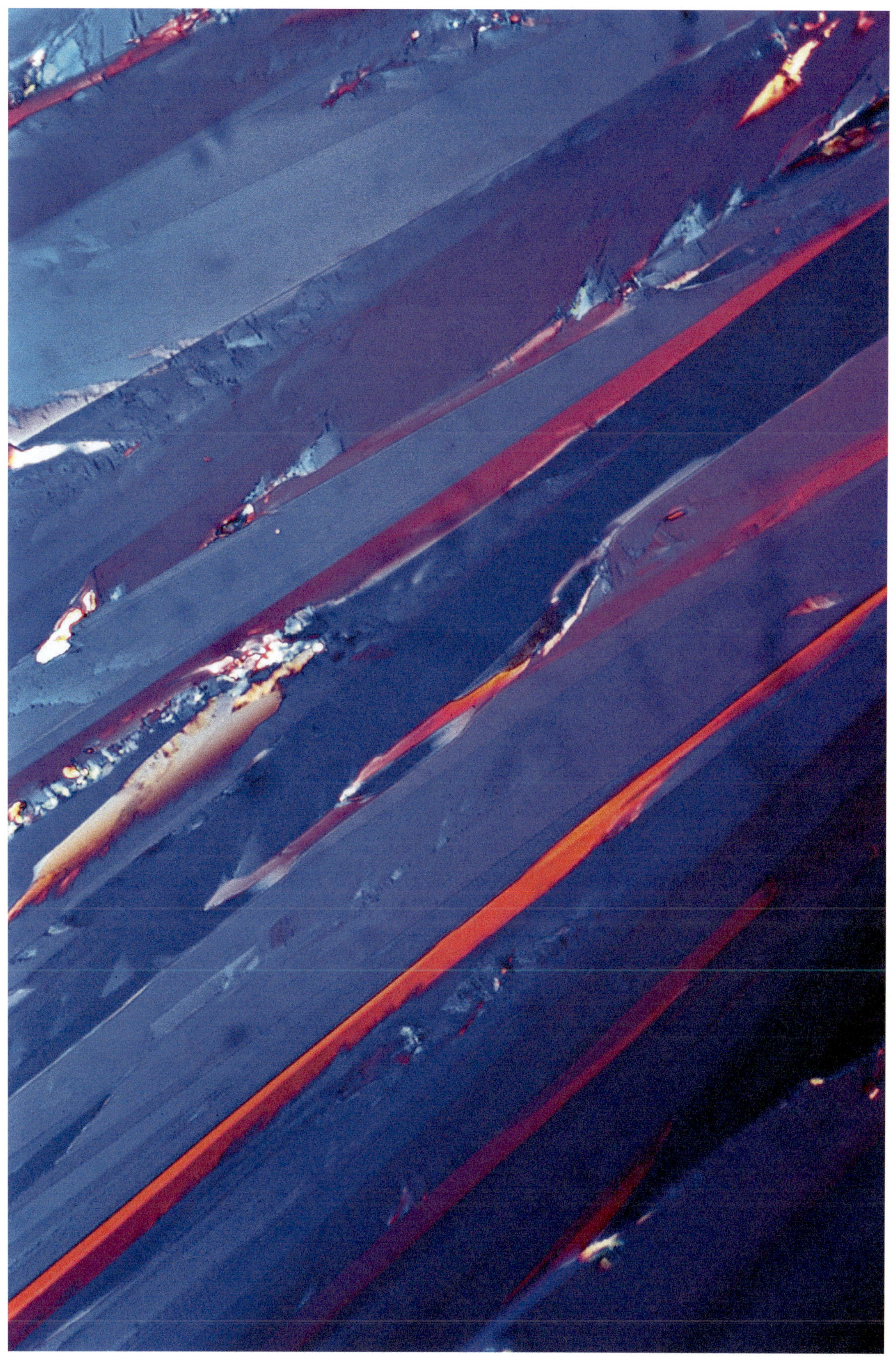

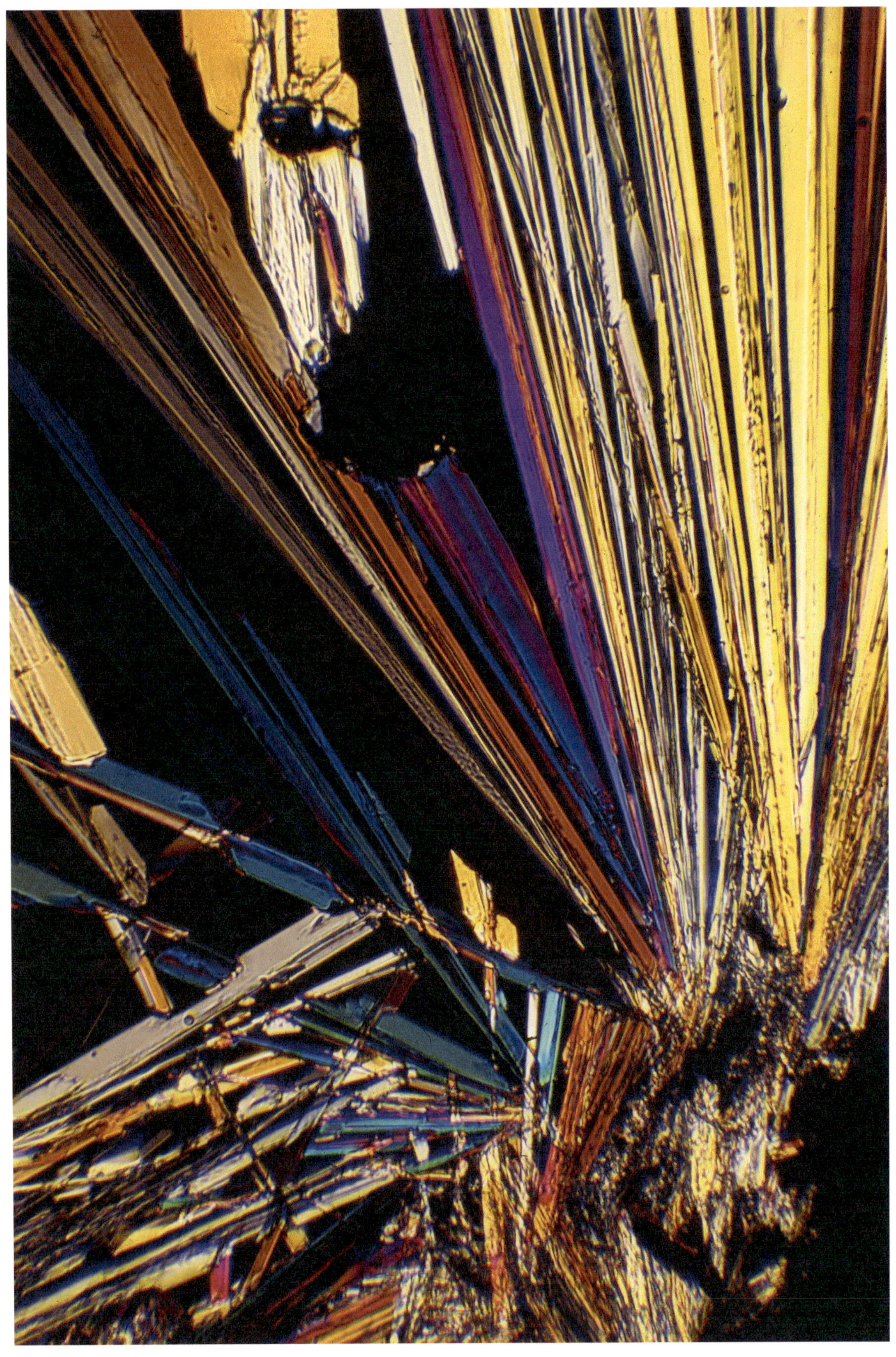

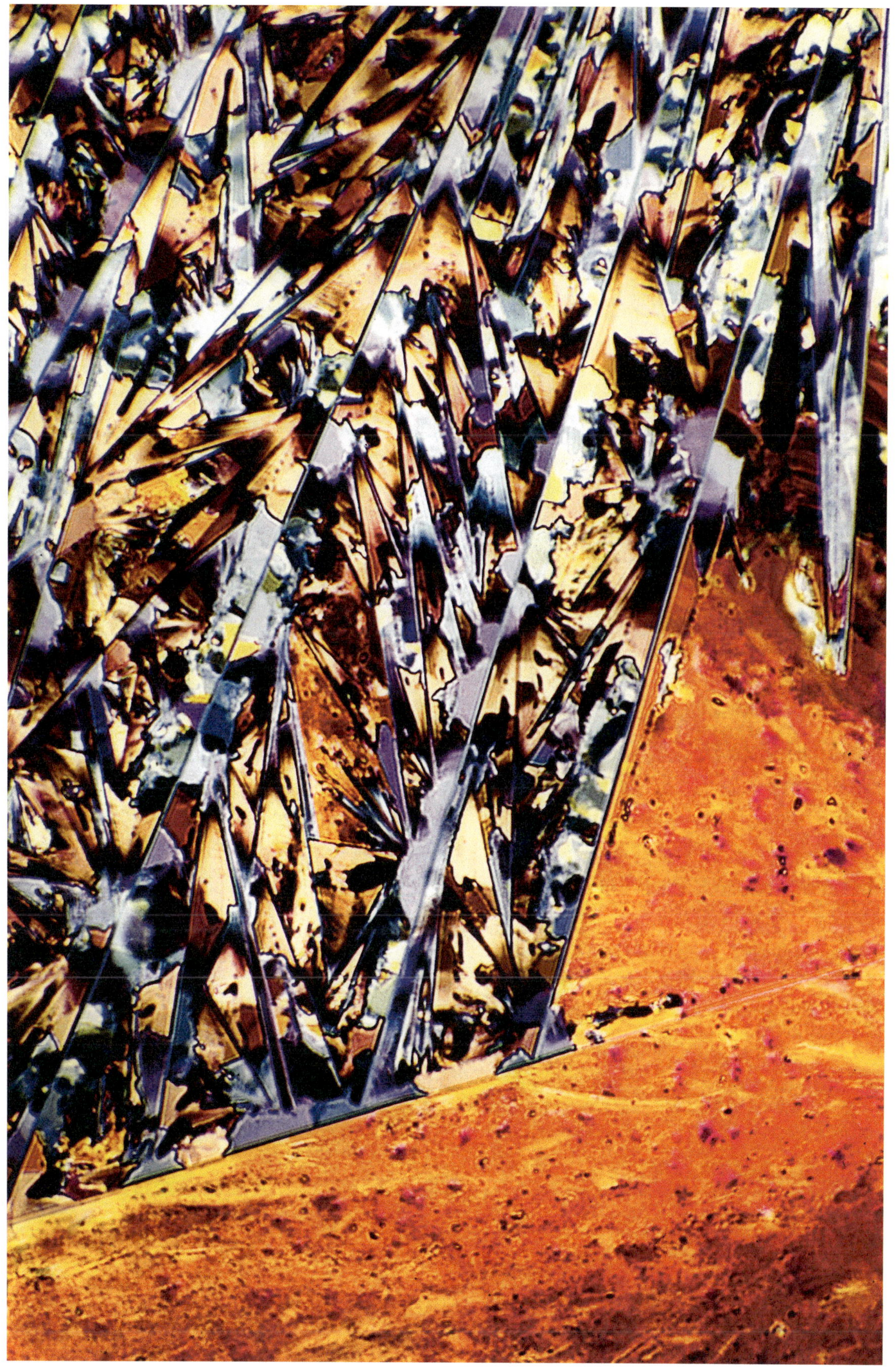

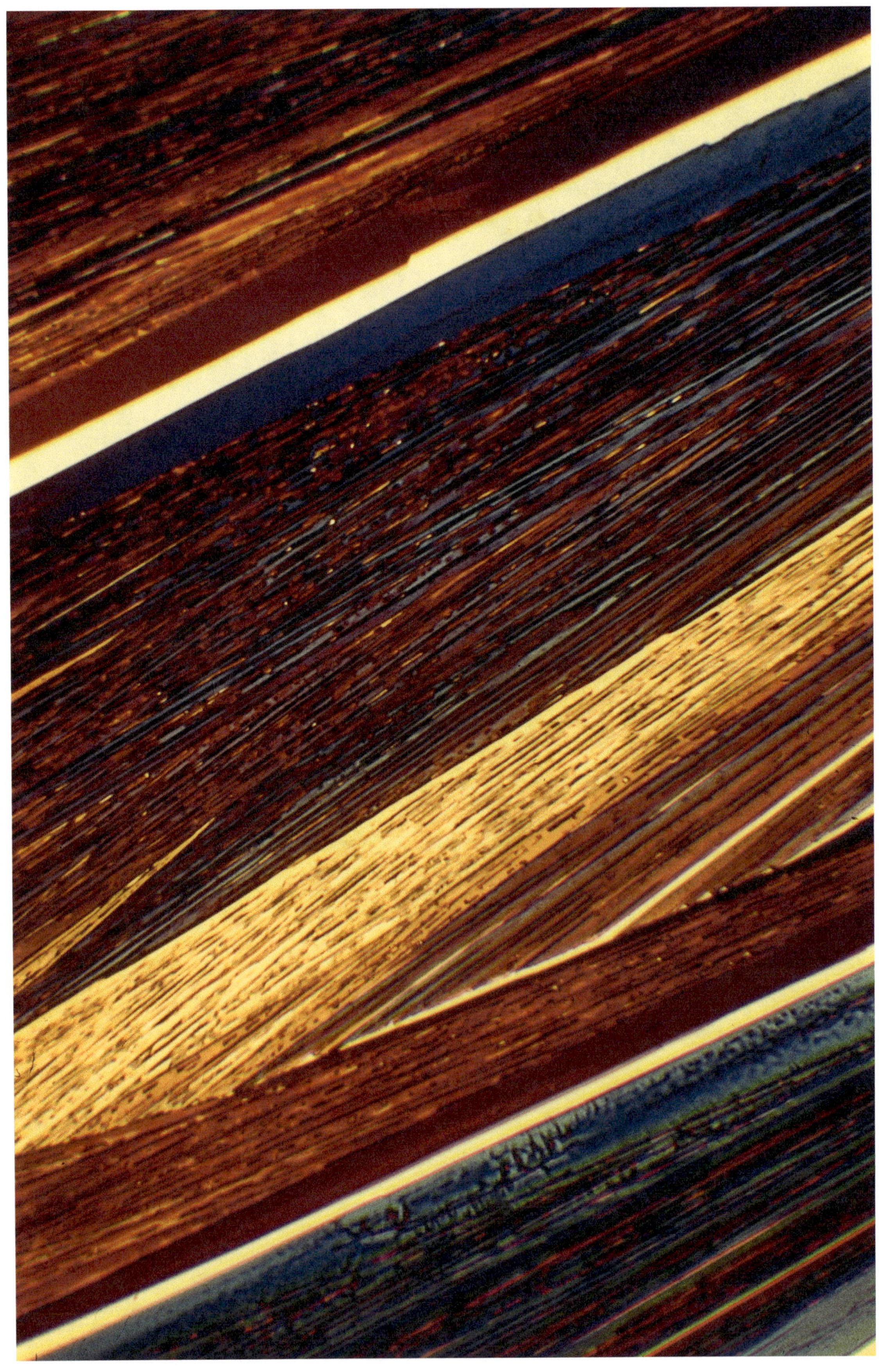

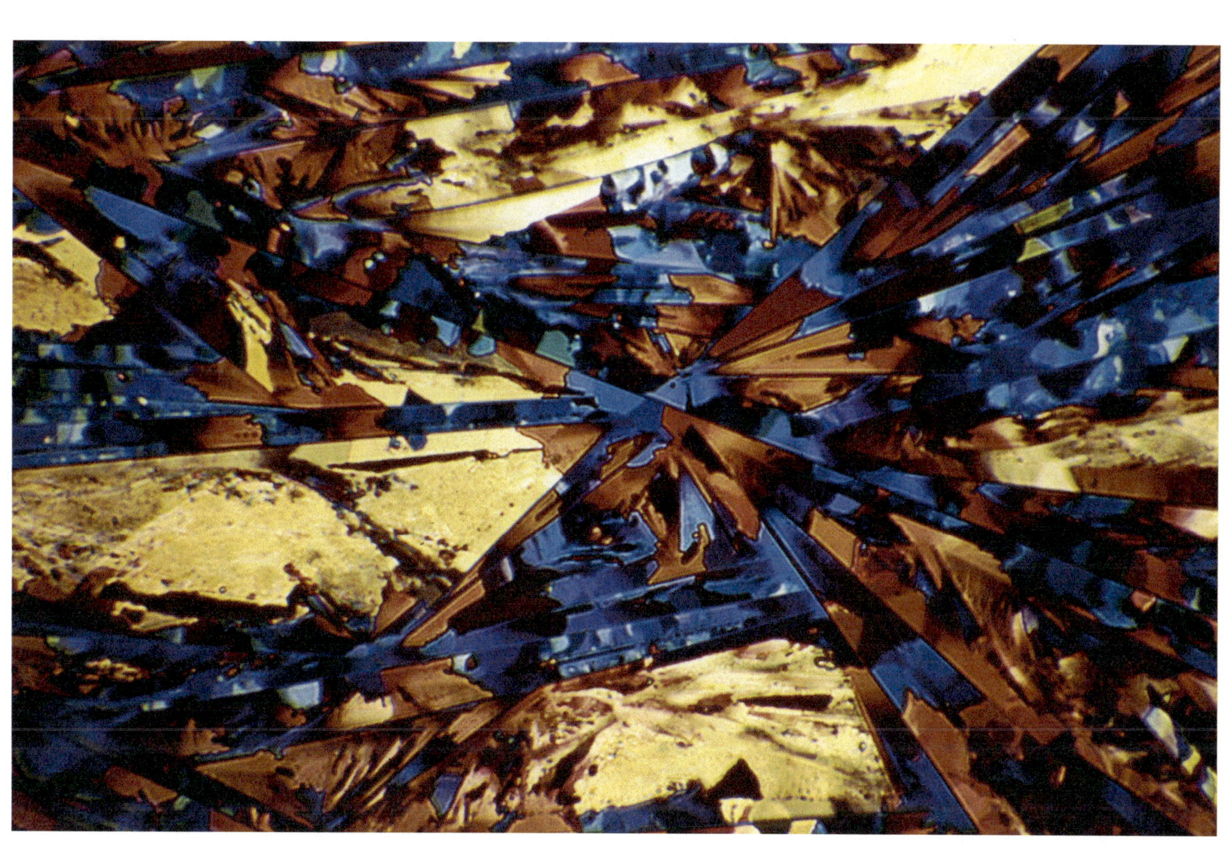

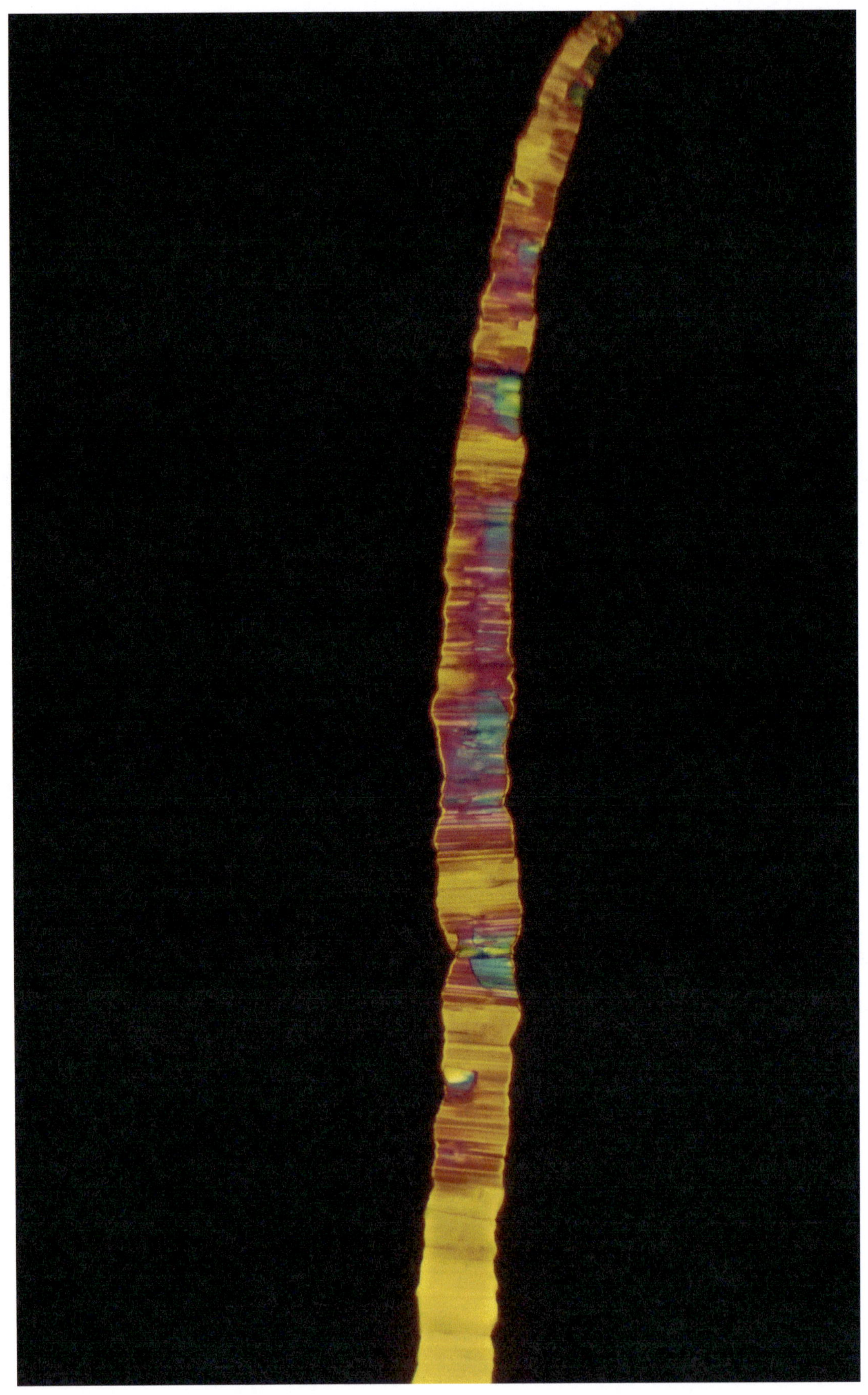

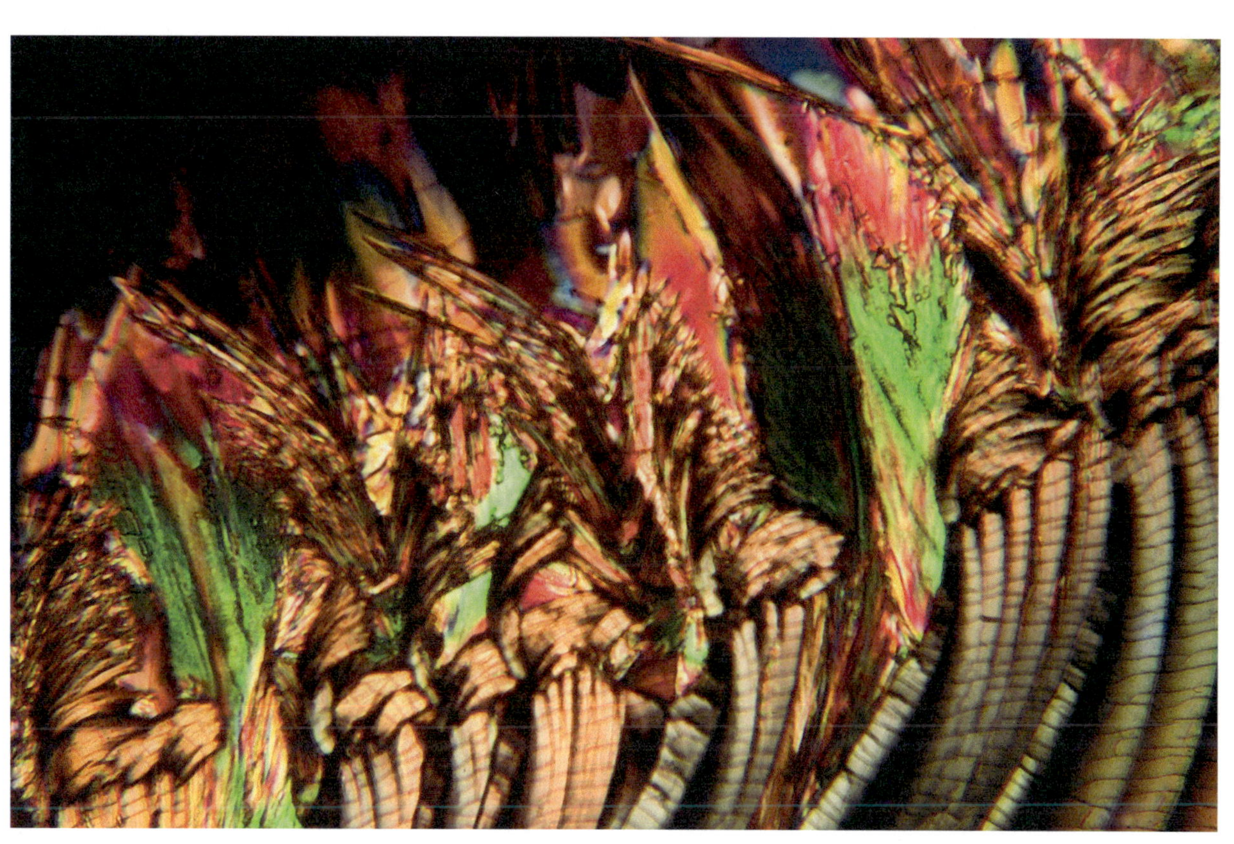

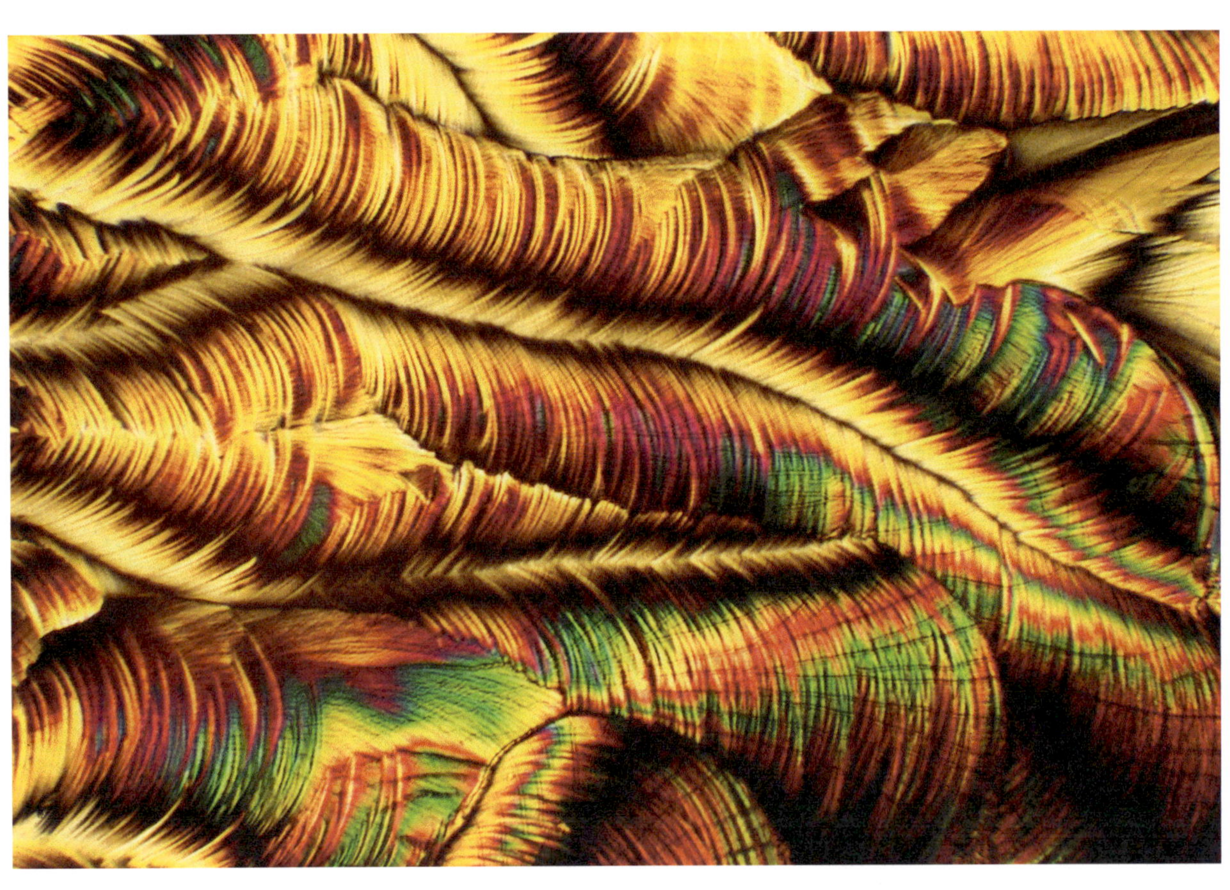

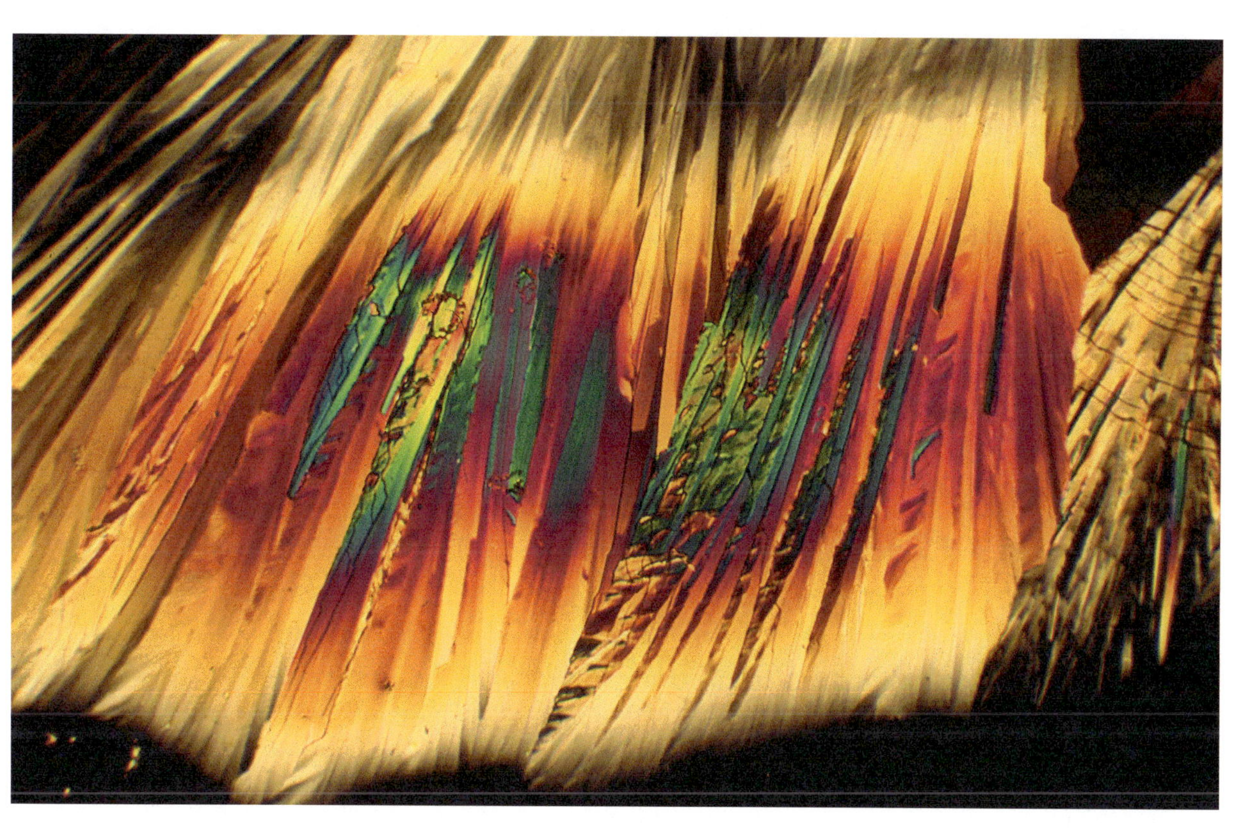

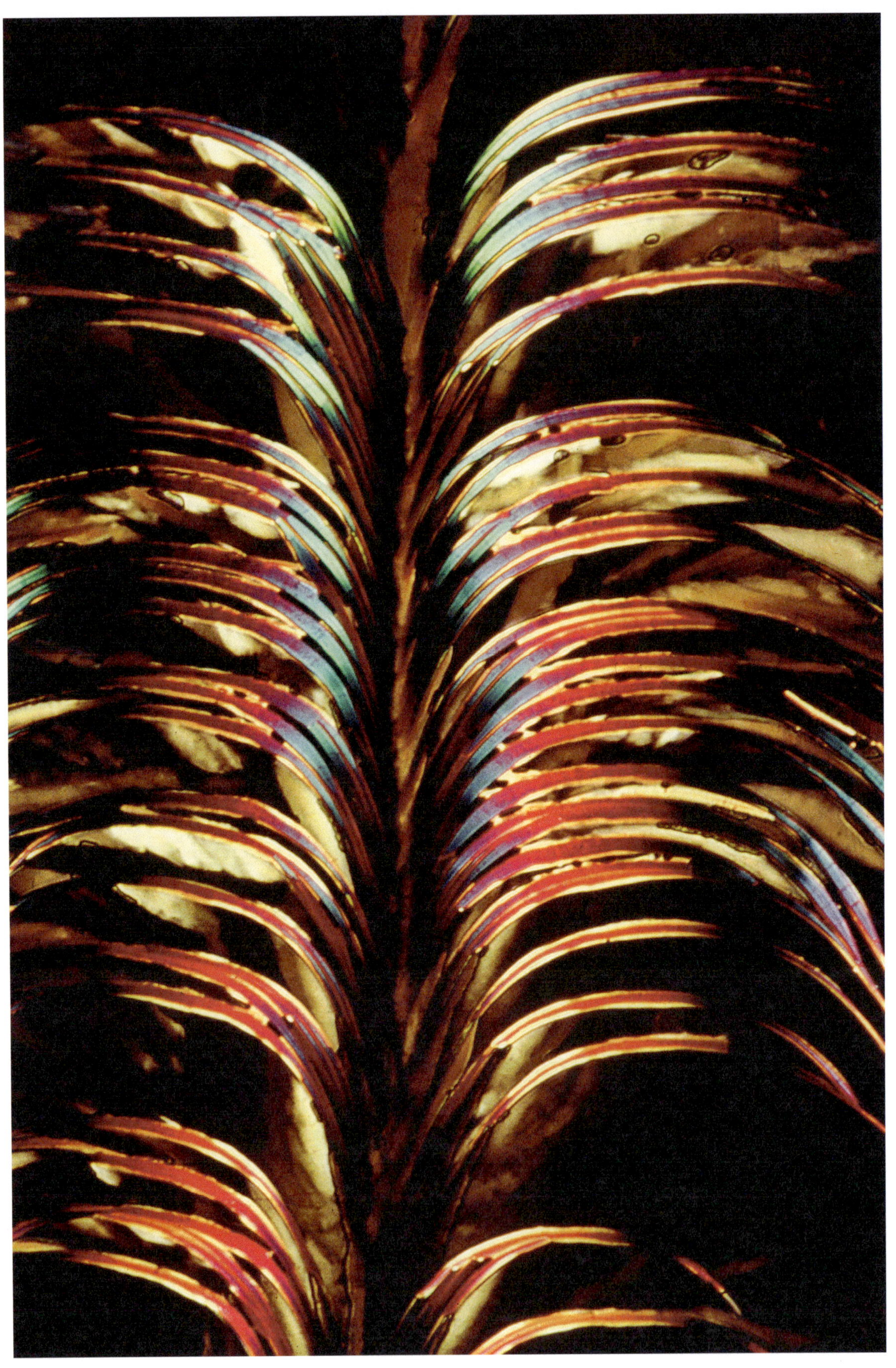

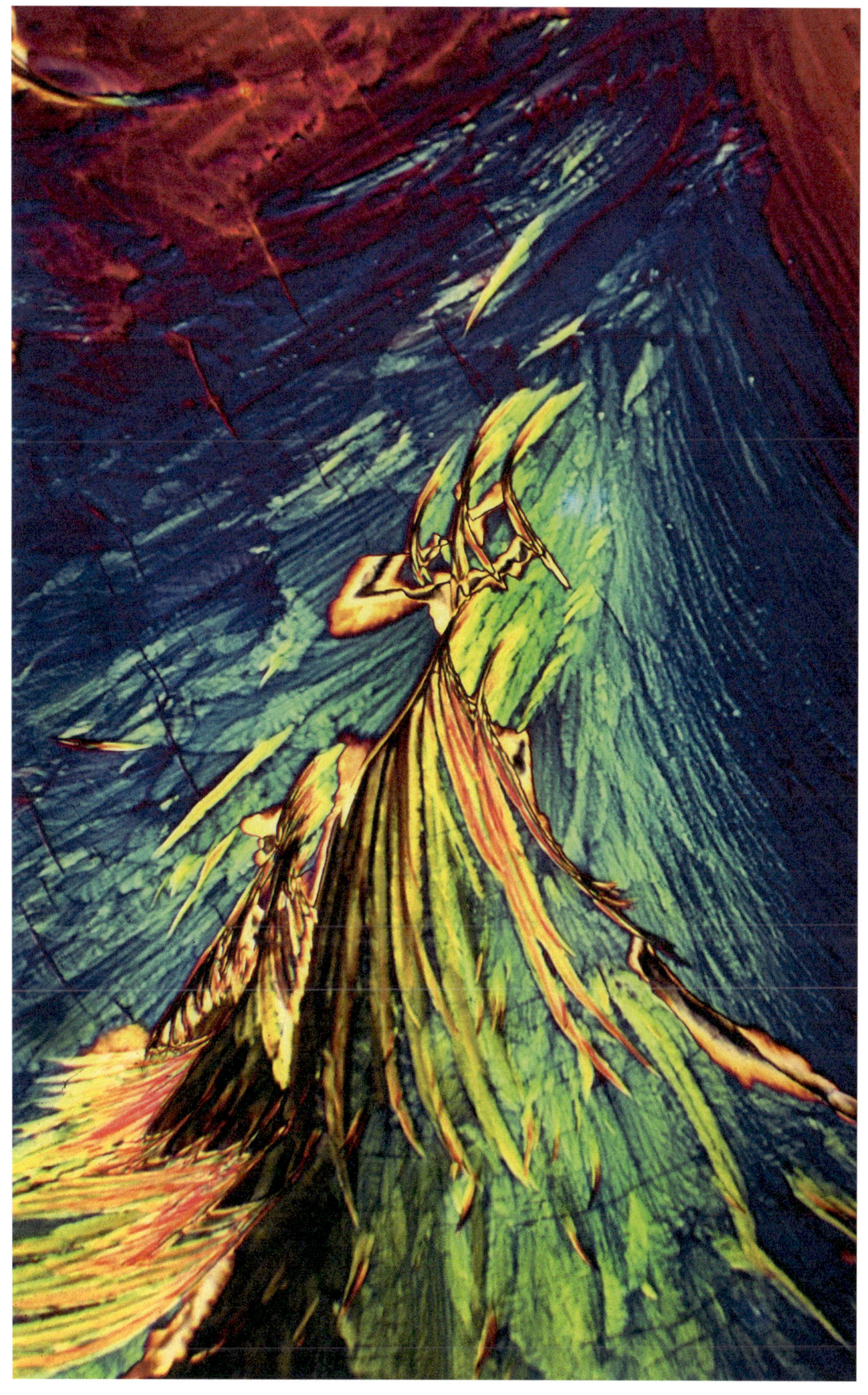

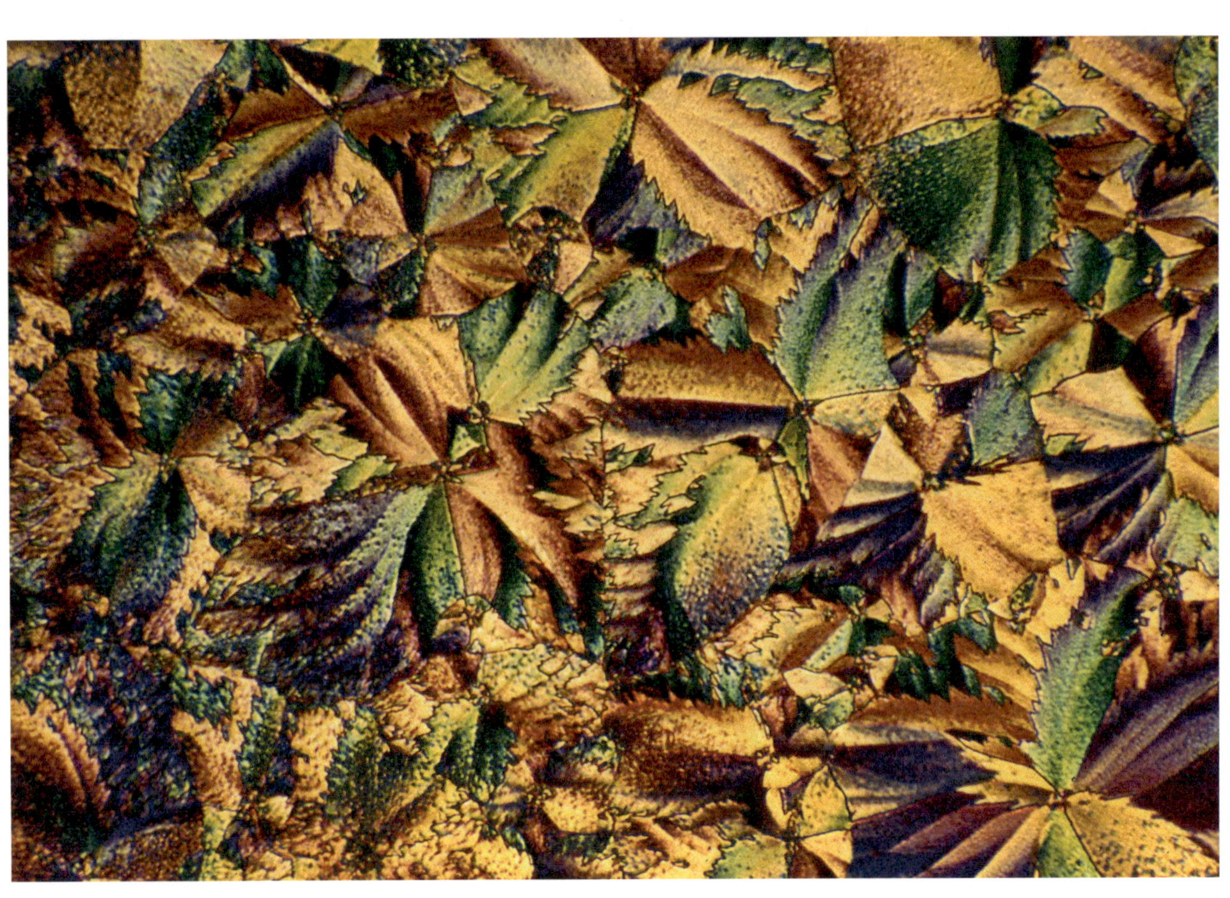

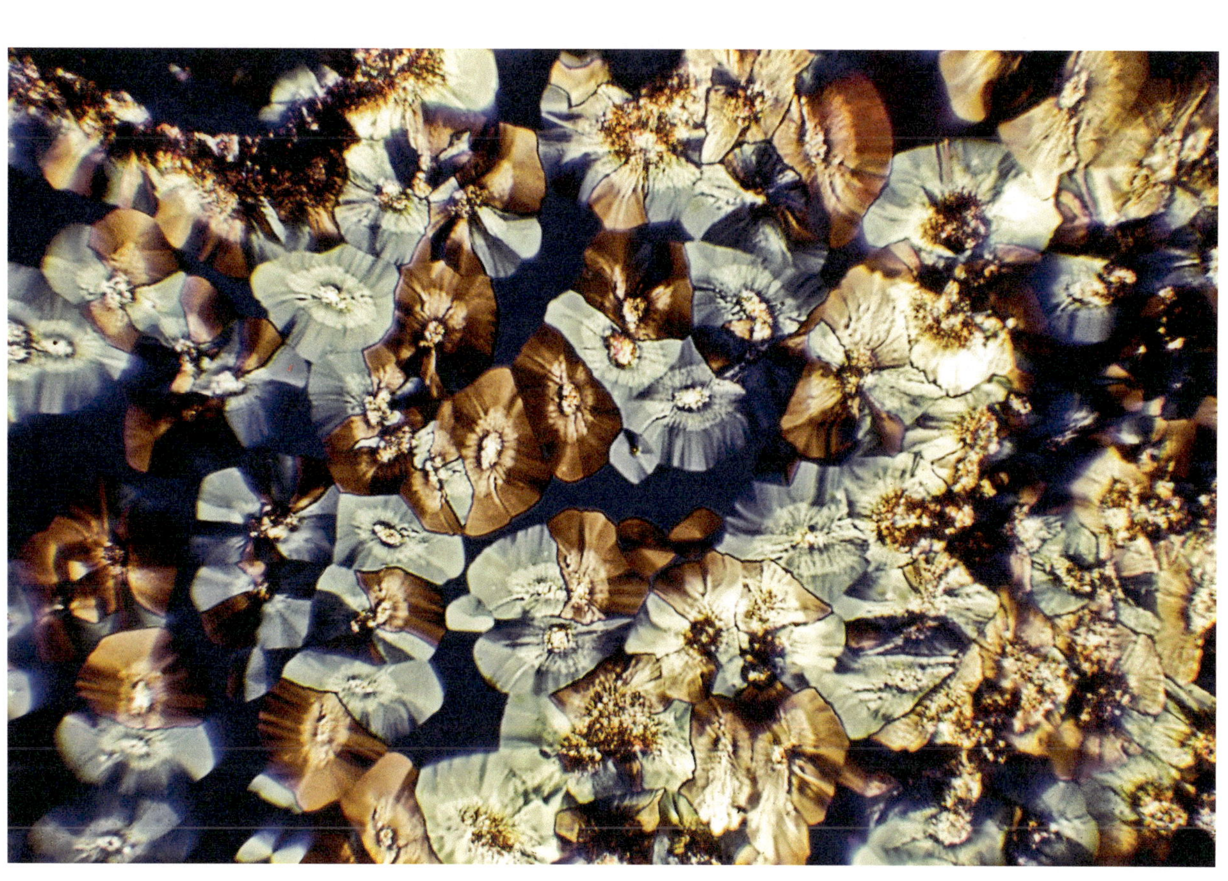

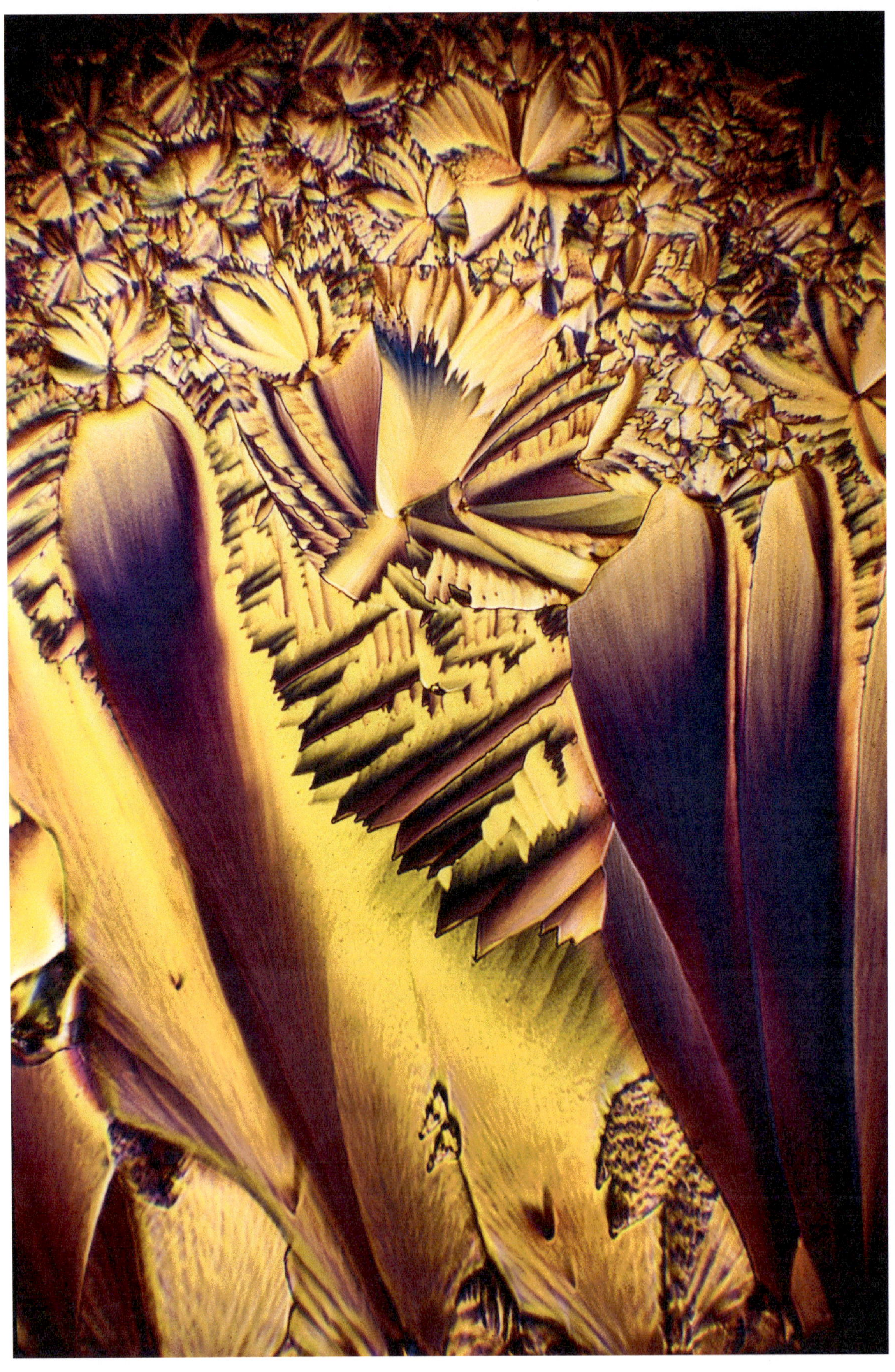

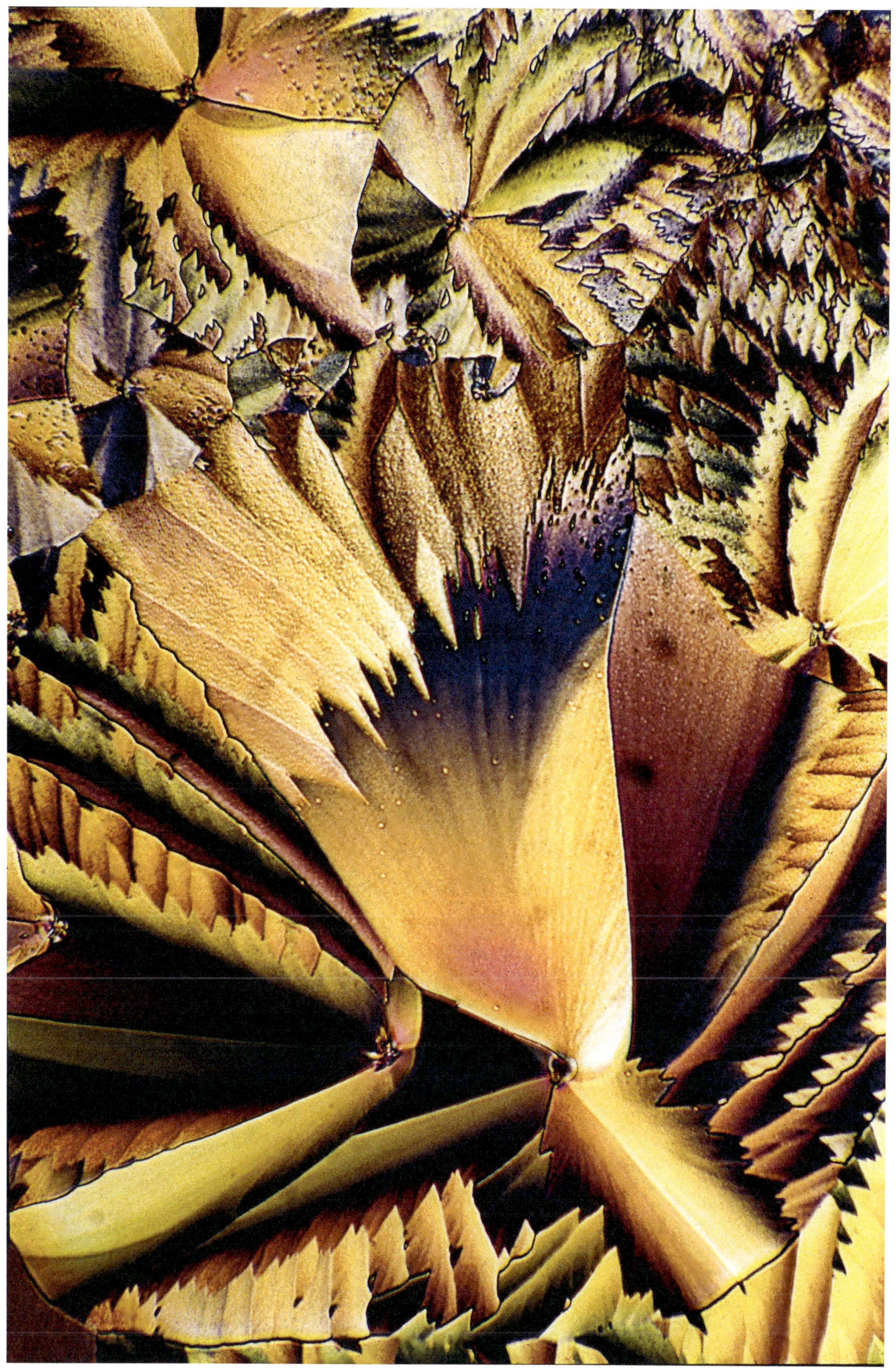

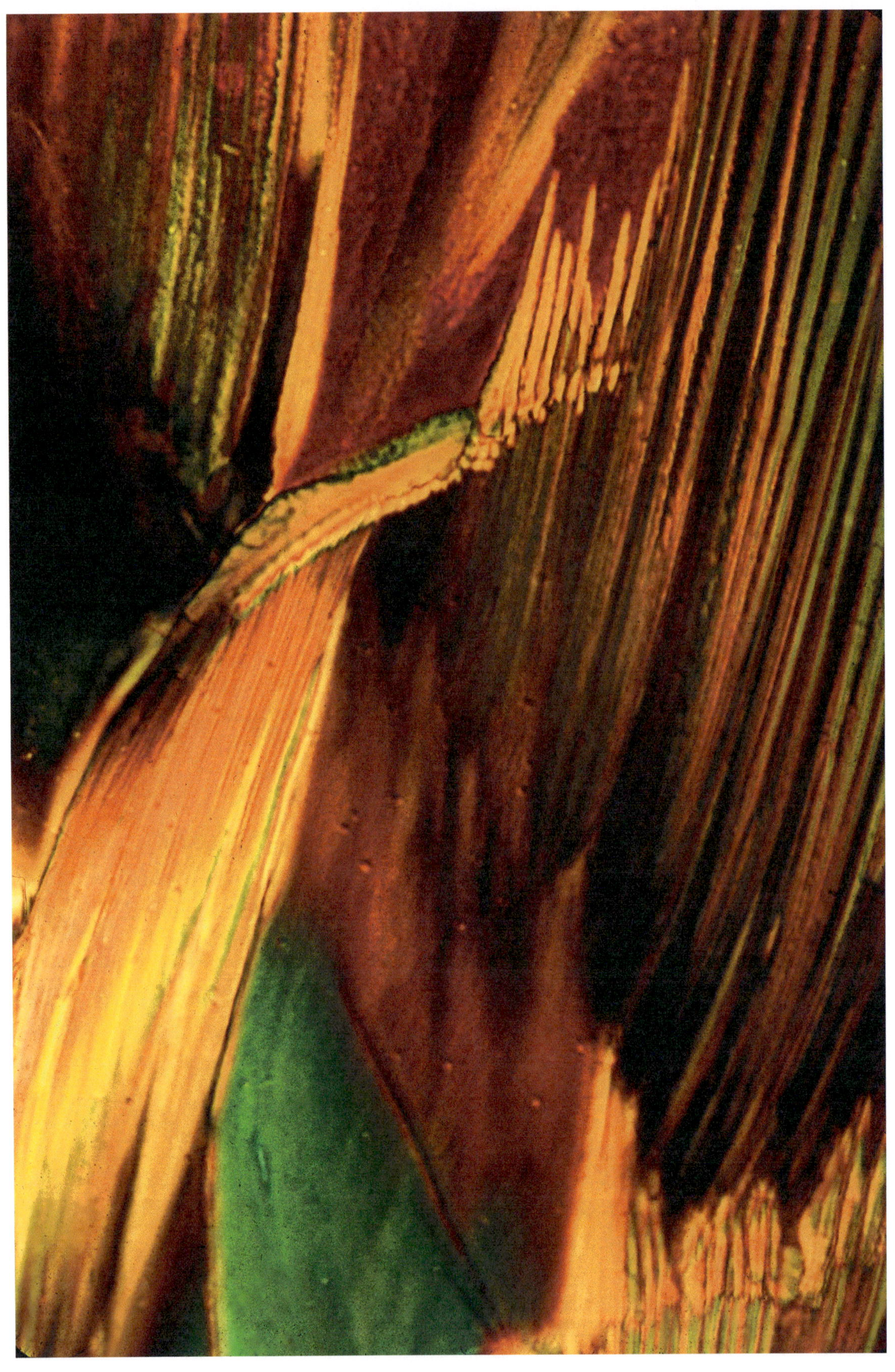

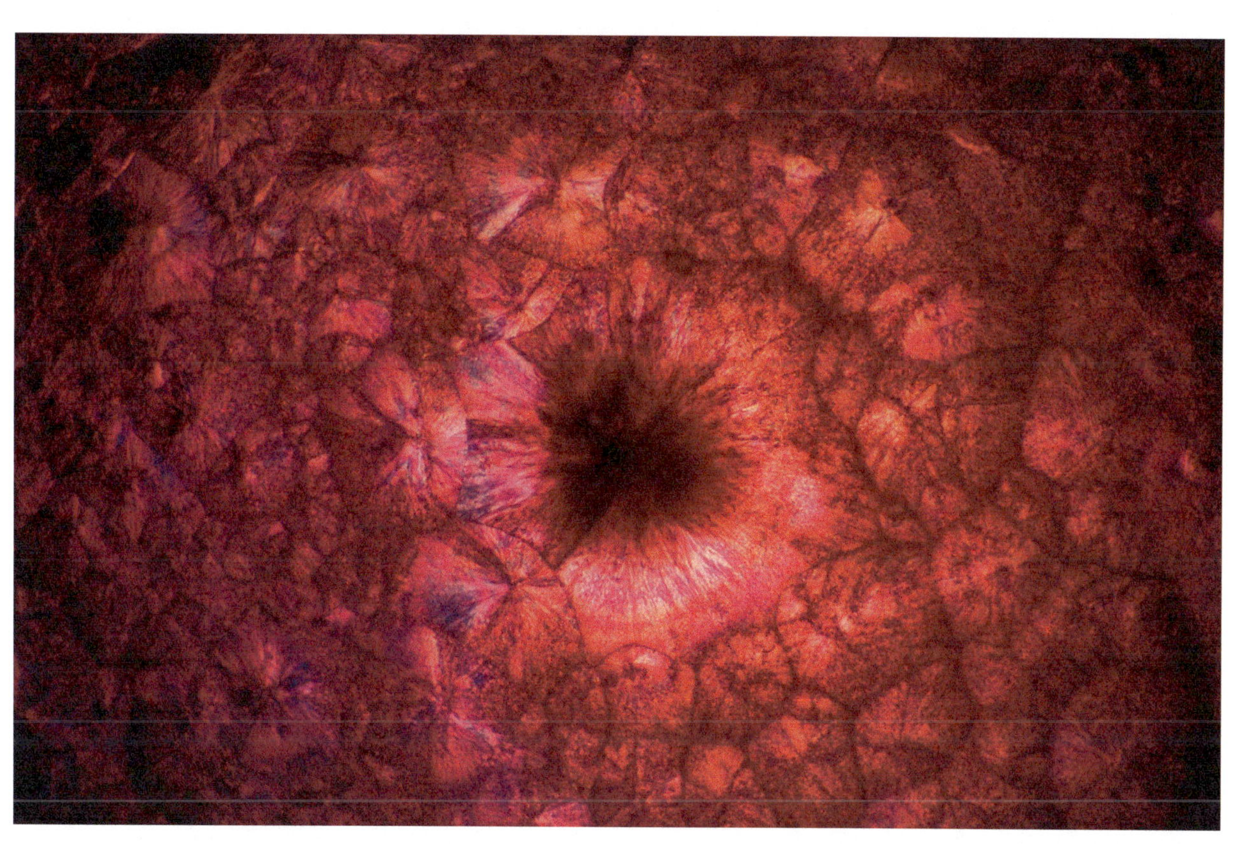

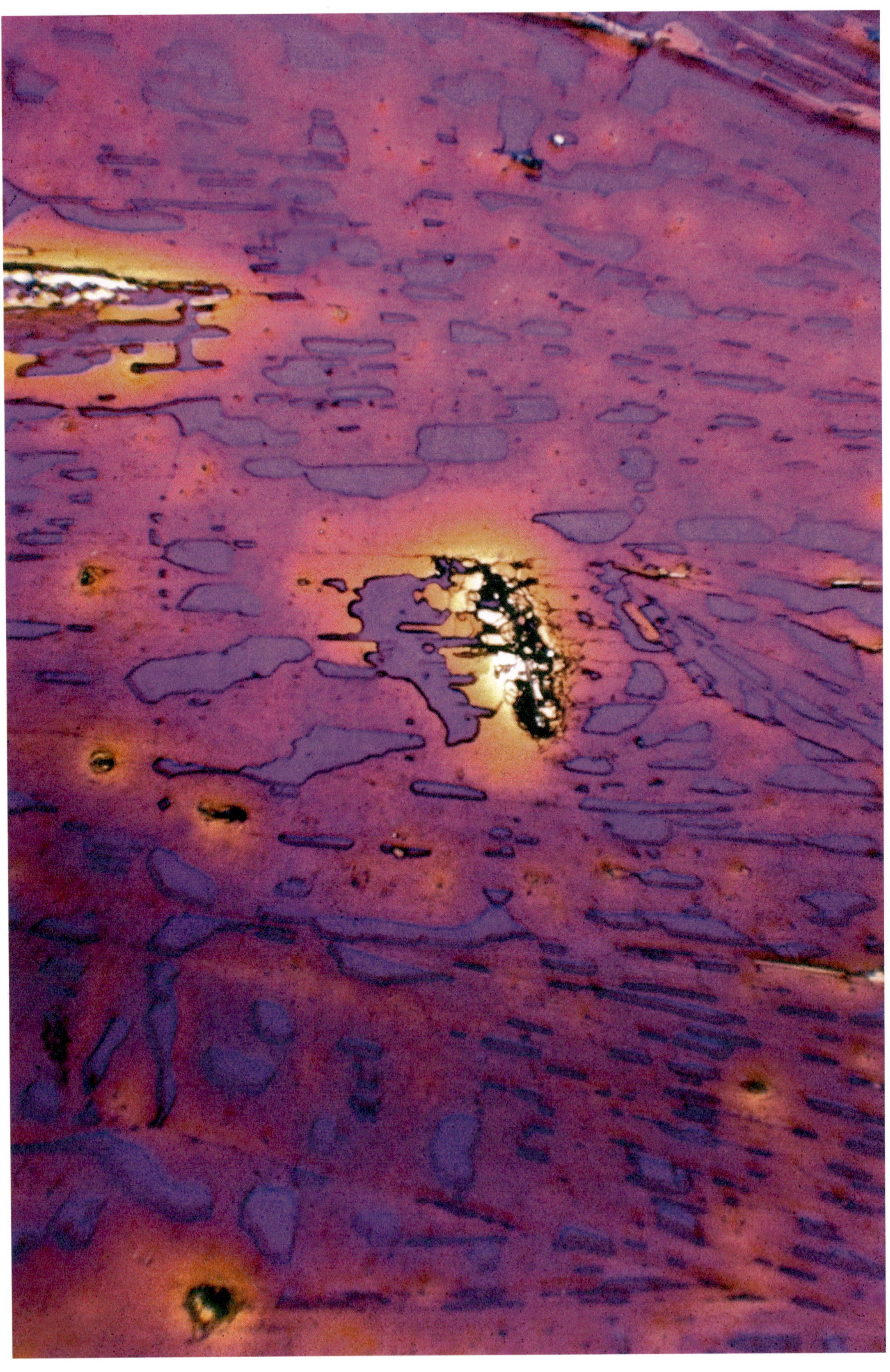

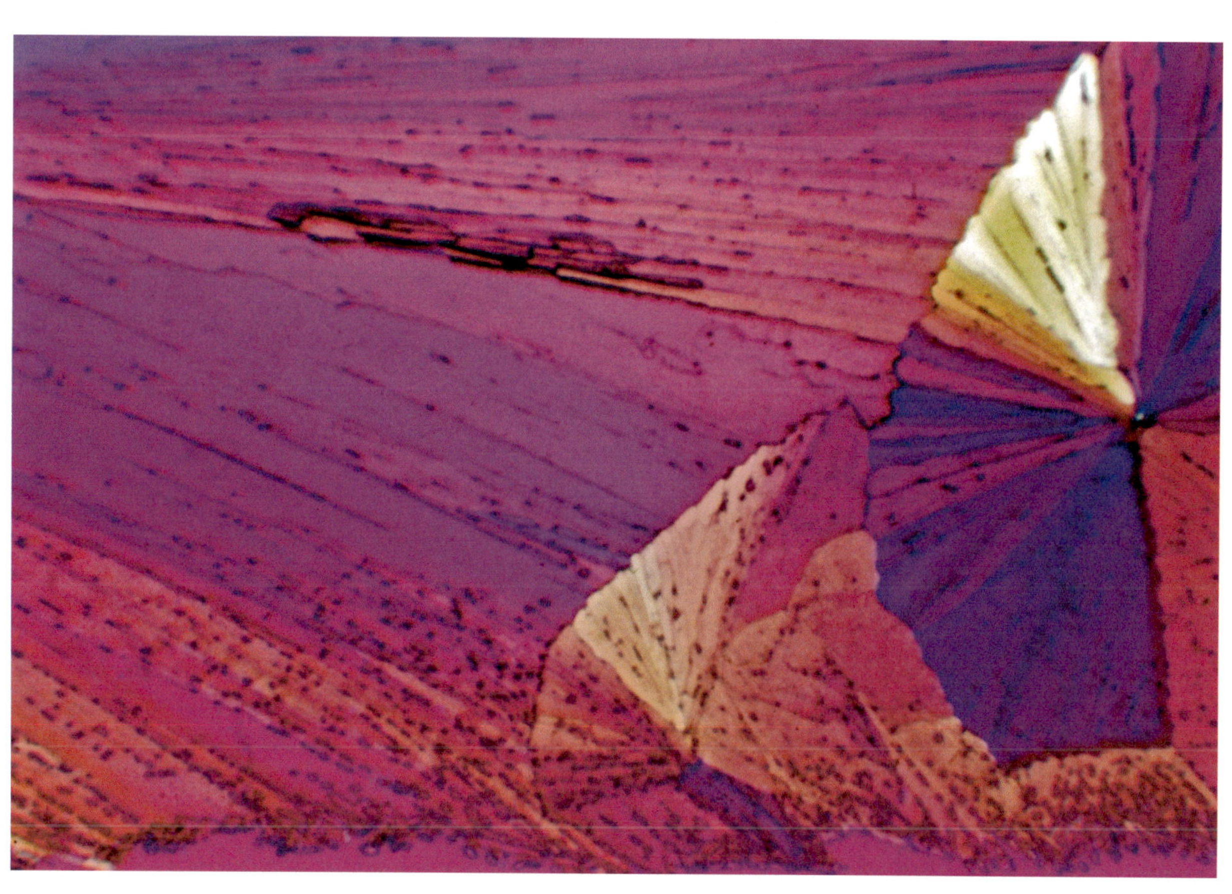

ABOUT THE AUTHOR

ARCHITECTURAL PHOTOGRAPHER and author Steven Brooke is a Fellow of the American Academy in Rome and winner of the American Institute of Architects National Award for Photogrpahy. He is the photographer of over 40 books on architecture and design, 10 of which he also wrote. Steven is on the facultyf of the University of Miami's School of Architecture. He conducts seminars and workshops and lectures regularly throughout the country. His work may be seen at www.stevenbrooke.com.

STEVEN BROOKE: SELECTED BOOKS

Views of Rome (Rizzoli)
Views of Jerusalem and the Holy Land (Rizzoli)
Miami Beach Deco (Rizzoli)
Deco Delights (Dutton)
Seaside (Pelican)
Addison Mizner (Rizzoli)
Houses of Philip Johnson (Abbeville)
Great Houses of the South (Rizzoli)
Historic Houses of Virginia (Rizzoli)
Great Houses of Florida (Rizzoli)
Carrére and Hastings: The Masterworks (Rizzoli)
Majesty of Natchez (Pelican)
Majesty of St. Augustine (Pelican)
Aqua: Miami Modern on the Beach (Rizzoli)
Savannah Style (Rizzoli)
Napa Valley Style (Rizzoli)
Sonoma Valley Style (Rizzoli)
Seaside Style (Rizzoli)
Historic Washington, Arkansas (Pelican)
Gardens of Florida (Pelican)
Louisiana Gardens (Pelican)
Sacred Journey (RDJ Publications)
The Reincarnation of Manuel Berzunza (Brooke Publications)
Miami: Mediterranean Dreams and Deco Delights (Rizzoli)
Florida Modern (Rizzoli)
Casa Florida (Rizzoli)
Dennis Jenkins Design (Brooke Publications)
Gunston Hall
The Hotel Ponce de Leon (Flagler College Press)
The Houses of Marc Corbiau
Vizcaya Museum and Gardens (U. Pennsylvania Press)